THE INDUSTRIAL GEOGRAPHY OF ISRAEL

In some ways, Israel's industrial geography is unique. The continuing Arab–Israeli conflict has been a primary force behind government intervention in settlement patterns, and has led to a major effort to disperse industry. The geopolitical situation has also encouraged a policy of attempted self-reliance, especially for defence purposes. These factors, combined with abundant human capital, have given Israeli high-technology industries a special place in the international division of labour. The absorption of waves of mass immigration has influenced industrial development. Rural industrialisation, mainly by the kibbutz (communal settlement) movement, is another unique feature.

The Industrial Geography of Israel presents a comprehensive overview of the industrial spatial development of Israel from the Ottoman era to present times, evaluating industrial dispersal policy, corporate geography, high-technology industries, entrepreneurship, and rural industrial development. The spatial development of Israeli industry is set within the broader context of Israel's political and economic development and of global economic change, as well as theories of industrial location and regional planning and development.

Yehuda Gradus holds the Harry Levy Chair in Geography and Regional Planning and is Head of the Centre for Regional Development Policy at Ben-Gurion University of the Negev, Israel. **Eran Razin** is Senior Lecturer in Geography at Hebrew University, Jerusalem, and **Shaul Krakover** is Senior Lecturer and Chairman of the Department of Geography and Environmental Development, also at Ben-Gurion University of the Negev.

THE INDUSTRIAL GEOGRAPHY OF ISRAEL

Yehuda Gradus, Eran Razin and Shaul Krakover

London and New York

First published 1993
by Routledge
11 New Fetter Lane, London EC4P 4EE

Simultaneously published in the USA and Canada
by Routledge
29 West 35th Street, New York, NY 10001

© 1993 Yehuda Gradus, Eran Razin and Shaul Krakover

Typeset in 10 on 12 point Garamond by
LaserScript, Mitcham, Surrey

Printed and bound in Great Britain by
Biddles Ltd, Guildford and King's Lynn

All rights reserved. No part of this book may be reprinted or
reproduced or utilized in any form or by any electronic,
mechanical, or other means, now known or hereafter
invented, including photocopying and recording, or in any
information storage or retrieval system, without permission in
writing from the publishers.

British Library Cataloguing in Publication Data
A catalogue record for this book is available from the British Library

Library of Congress Cataloging in Publication Data
Gradus, Y., 1942–
The industrial geography of Israel/by Yehuda Gradus, Eran Razin
and Shaul Krakover.
p. cm.
Includes bibliographical references and index.
ISBN 0–415–02156–1
1. Israel – Industries – Location. 2. Industry and state – Israel.
I. Razin, Eran. II. Krakover, Shaul, 1947– . III. Title.
HC415.25.Z9D54
338.6é042é095694 – dc20   92-37810
CIP

ISBN 0–415–02156–1

CONTENTS

List of figures	ix
List of tables	xi
Preface	xiii

Part I The industrial geography of Israel: An introduction

1 GENERAL BACKGROUND	1
Introduction	1
The purpose and structure of the book	6
2 ISRAELI INDUSTRY: BACKGROUND AND ROLE IN THE ECONOMY	8
Global shifts and economic and political change in Israel	8
National comparative advantage	13
Major attributes of Israeli industry and its role in the economy	15
An international comparative perspective	22

Part II The evolution of the industrial geography of Israel until 1973

3 INDUSTRIAL EVOLUTION IN ISRAEL: GENERAL ARGUMENTS	27
4 THE PRE-STATEHOOD ROOTS	34
An overview	34
End of the Ottoman era – traditional local specialisations and seeds of modern industry	36

CONTENTS

*The British Mandate era – industrial development in a
dual Jewish–Arab economy* 38
Spatial transformation during the British Mandate era 45

5 EARLY CRYSTALLISATION OF THE ISRAELI SPATIAL
INDUSTRIALISATION POLICY, 1948–55 51

6 THE GREAT PUSH FORWARD: INDUSTRIALISATION
OF THE DEVELOPMENT TOWNS, 1956–67 56
Realisation of preconditions for industrial dispersal 56
*The golden years of industrial dispersal – an
introduction to the Sapir era* 58
Early failure – the case of the diamond industry 63
Spatial dispersal led by the textiles sector 66
Implications for industrialisation of the large cities 68

7 THE POST-1967 CROSSROADS 72

**Part III The industrial geography of Israel in a
period of economic stagnation**

8 CHANGING REALITIES OF THE 1970s AND 1980s 79

9 MEASURING THE EFFECTIVENESS OF INDUSTRIAL
DISPERSAL POLICY: A QUANTITATIVE ASSESSMENT 83
Level of industrialisation 83
Level of diversification 86

10 STAGNATION OF THE INDUSTRIAL BASE IN THE
PERIPHERY AND THE CASE FOR DIVERSIFICATION 91

11 INERTIA AND INCREMENTAL CHANGE: AN
EVALUATION OF THE INCENTIVES FOR INDUSTRIAL
DISPERSAL 98
Evolution, attributes, and distortions of policy means 98
The need for revision 103

12 CORPORATE GEOGRAPHY AND THE CRISIS IN THE
FEDERATION OF LABOUR ENTERPRISES 113
Location of headquarters of the large industrial firms 114
The spatial organisation of the largest multiplant firms 116
*The role of ownership characteristics in industrial
    development in the periphery* 119
The crisis in the Federation of Labour enterprises 121

CONTENTS

13	THE EMERGING GEOGRAPHY OF HIGH-TECHNOLOGY INDUSTRIES	124
	The growth of Israel's high-technology industries: an overview	124
	The evolution of major high-technology spatial clusters	137
	Attempts at dispersal	147
	Implications of the crisis of the 1980s – the mixed blessing of unselective dispersal	152
14	THE RE-EMERGENCE OF LOCAL DEVELOPMENT STRATEGIES	157
	Factors encouraging the emergence of local development policies in Israel	157
	Competing strategies	158
	The spatial implications	165
15	THE LOCAL ENTREPRENEURSHIP OPTION	169
	Introduction	169
	Trends and spatial variations	170
	Shifting attitudes towards entrepreneurship – from an evil to a blessing	172
	Promoting entrepreneurship and the development of the periphery – early attempts	174
	The take-off – a major route for absorbing immigrants or a late adaptation of the policy of the 1980s?	177

Part IV   Rural industrialisation in Israel

16	INDUSTRIALISATION IN RURAL ISRAEL: AN INTRODUCTION	183
17	THE UNIQUE CASE OF THE KIBBUTZ	186
	The kibbutz and its basic ideology	186
	Industrialisation in the kibbutz	191
18	THE RURAL MOSHAV INDUSTRIALISATION PROCESS	210
	The organisation of the moshav	210
	Industrialisation of the moshav	213
19	INDUSTRIALISATION IN THE ARAB SECTOR	217
	Rural industrialisation in Arab villages	217

Part V Postscript

20 ISRAEL'S INDUSTRIAL GEOGRAPHY IN THE 1990s 225
*Arabs and Jews in the occupied territories – aspects of
industrialisation* 226
Immigration and spatial change 228
Conclusions and future scenarios 230

Bibliography 235
Index 251

FIGURES

1.1	Israel – regions and topography	5
2.1	Israeli industry – production indices, 1955–90	15
2.2	Israeli industry – number of employed persons, 1954–90	16
2.3	The share of industry in the Israeli labour force and net domestic product, 1954–90	17
2.4	Israeli industry – employment indices for selected branches, 1952–90	19
2.5	Domestic product of Israeli industry by major branch, 1965/6–88	20
2.6	Israel's industrial exports by major branch, 1970–90	23
3.1	Scheme for identification of stages in the development of Israel's spatial industrialisation policy	31
4.1	Industrial specialisation in the large urban centres of Palestine during the late Ottoman period	37
4.2	Palestine – spatial distribution of Jewish-owned industrial establishments, 1930–43	48
5.1	Israel – employed persons in industry by region and major branch, 1952	53
6.1	Israel – district and subdistrict boundaries	61
6.2	Israel – percentage employed in peripheral districts for selected industrial branches, 1955–88	62
9.1	Levels of industrial employment concentrations, 1965–87	84
9.2	Changes in the level of industrial employment concentration, 1965–87	87
9.3	Changes in the rank order of subdistricts in terms of industrial employment diversification, 1965–87	89

ix

FIGURES

11.1	Israeli development zones according to the Law for Encouraging Capital Investment: 1967, 1972, 1977, 1988, 1991	104
13.1	The spatial distribution of high-technology employment, 1969/70	138
13.2	The spatial distribution of high-technology employment by industry, 1982/3	140
13.3	The spatial distribution of high-technology employment in the Tel Aviv metropolitan area by industry, 1982/3	143
14.1	Core, semi-periphery and periphery in the Israel of the 1980s – a preliminary sketch	167
17.1	Comparison between kibbutz industry and that of the whole of Israel by major branches	200

TABLES

2.1	General overview of external economic and political conditions, and shifts in the Israeli spatial policy	10
2.2	Employed persons in Israeli industry by selected branches, 1952–1990	18
2.3	Israeli industrial exports, 1949–90	22
3.1	A summary of stages in the development of Israel's spatial industrialisation policy	32
4.1	Palestine – employed persons in industry and size of average plant, 1928	40
4.2	Palestine – employed persons in the Jewish industrial sector, 1921–43	41
4.3	Palestine – Jewish sector and Arab sector industrial product, 1922–47	43
4.4	Palestine – employed persons in industry and size of average plant, 1942	44
4.5	Palestine – concentration of industry in the four largest urban centres, 1928	46
6.1	Israel – employed persons in industry by district	59
6.2	The Israeli diamond industry, 1946–72	64
6.3	Employed persons in plants owned by Koor Industries, 1954–80	70
10.1	Plant closures in Israel's development towns	92
11.1	Israeli industry – indices of capital intensity, 1982	101
12.1	Location of head offices of the largest industrial firms in Israel, 1981–8	115
12.2	Employees of large industrial firms located in development regions	118
12.3	Israel – employed persons in industry by sector of ownership, 1965–87	122

TABLES

13.1	Firms in selected high-technology manufacturing industries, 1990	141
15.1	Israel's urban population, self-employed and self-employed in manufacturing, 1961–83	171
17.1	Development of manufacturing plants in kibbutzim, 1946–90	195
17.2	Distribution of kibbutz industrial activities, 1970, 1979, 1988	196
17.3	Changes in plant size, 1970, 1979, 1988	197
17.4	Weight of the kibbutz in the Israeli manufacturing sector, 1960, 1972, 1982	199
17.5	Examples of product types by branch of industry owned by kibbutzim	201
18.1	Examples of selected moshav industrial plants in the Galilee	215
20.1	Israel – approved projects for industrial investment in manufacturing, 1984–91	227

PREFACE

The Industrial Geography of Israel is the first book to present a comprehensive overview of economic industrial development from a spatial perspective. The book is based on an integration of the authors' previous research and additional material gathered especially for this volume.

Even though we formulated the structure and overall approach of the text together, each of us has taken special responsibility for different aspects of the book. Yehuda Gradus was mainly concerned with rural industrial development in the kibbutz, moshav and Arab sectors and for the introductory chapter. Eran Razin took primary responsibility for the post-1948 evolutionary development of Israeli industry, evaluating the incentives for the industrial dispersal policy, corporate geography, the emergence of high-technology industries, and entrepreneurship and local development. He is also responsible for the Postscript. Shaul Krakover was responsible for the quantitative assessment of the effectiveness of the Israeli industrial dispersal policy, and for the Ottoman and British Mandate periods, and the chapter on the role of industry in the Israeli economy. Needless to say, all of us are jointly responsible for the book's style and content.

Figures were drafted by Ms Anat Bloch of the Cartographic Unit of the Department of Geography, The Hebrew University of Jerusalem.

We are grateful for the substantial English editorial, clerical and secretarial assistance of Ms Catherine T. Logan, who is editor and copy-editor of publications of the Hubert H. Humphrey Institute for Social Ecology, Ben-Gurion University of the Negev, and Managing Editor of *Israel Social Science Research*

Special appreciation is also expressed to Daniel Felsenstein of The Hebrew University, who reviewed the entire manuscript and provided us with very helpful comments and suggestions. Peter Sowden of Croom Helm and later Routledge, and Tristan Palmer of Routledge, deserve special thanks for their patience and encouragement.

Yehuda Gradus and Shaul Krakover,
Ben-Gurion University of the Negev, Beer-Sheva
Eran Razin,
The Hebrew University, Jerusalem
1992

Part I

THE INDUSTRIAL GEOGRAPHY OF ISRAEL: AN INTRODUCTION

1
GENERAL BACKGROUND

INTRODUCTION

The industrial geography of Israel presents an apparent paradox – while exhibiting many highly unique features, it none the less provides valuable insights into the study and theory of industrial geography in general.

The spatial organisation and structure of industry in any country is a reflection of a number of factors: the nature of its economy; its human and natural resources; geographical features, such as geopolitical status, which affect its role in the global economy; and political and ideological characteristics, as well as its cultural values and ethics. In this respect, Israel is no exception. With a population of some 5 million, Israel is a small country with an open economy and a scarce endowment of natural resources. It is, however, rich in highly educated and motivated human resources.

The major distinguishing features of Israel's space economy are closely linked to its existence as a Jewish state in a hostile environment. This unique geopolitical status has had a major impact on many aspects of life in the country, including industrial activity and its geographical distribution. The continuing Jewish/Arab conflict, accompanied by uncertainty as to Israel's future boundaries, has been a primary force behind intervention in settlement patterns by successive governments, particularly within the framework of a population dispersal policy. This also implies a major effort to disperse industry.

Uncertainty with respect to Israel's reliance on imports of strategic goods from more or less friendly nations has facilitated a policy of attempted self-reliance, or import substitution – especially for defence purposes. This has influenced the development, structure, and spatial distribution of Israeli industry, resulting in industrial activity that is to a large extent dependent on defence production, including a sophisticated aircraft sector, unusual for such a small country. Thus, high-technology industries have emerged primarily because of the need to develop sophisticated defence products. In addition, such industries best utilise available human resources and are competitive in export markets, despite the limitations imposed by the hostility of Israel's surrounding neighbours and the Islamic world in general, and by the Arab trade boycott in particular.

It can be seen that there has been a persistent dilemma in Israel's space economy between the needs for security and defence on the one hand, calling for (among other things) industrial dispersal, and, on the other, economic efficiency, requiring agglomeration and concentration of economic activities in metropolitan areas. Nevertheless, although a number of special features have had a unique influence on Israel's space economy, the nation's industrial development is not divorced from the general theories, concepts, and the body of knowledge pertaining to the study of industrial geography. For instance, despite Israel's small size, we can discern phenomena similar to those in larger countries, such as core/periphery dichotomies, spatial division of labour, the product cycle phenomena, and a developed system of incentives. The Israeli case may therefore provide new general insights into the study of industrial geography.

Immigration and ideology

Large-scale Jewish immigration and settlement schemes present another unique feature of the Israeli case. Indeed, Israel is a country whose very *raison d'être* has been the 'ingathering of the exiles'. The settlement patterns and initial industrial structure were mainly laid down by and for successive waves of immigrants coming from both developed and developing countries from about 1880 onwards. Israel's comparative advantage in human capital is one consequence of this process. A special relationship with world Jewry (an element that has influenced the country's economic development) is another. Among other factors, these have led the country to a special role in the global economy.

It should be noted, however, that whereas most of the pre-state immigration came from Western countries, the 1950s and 1960s saw major waves of immigration from the Third World, principally

Jewish communities from Arab countries, who then confronted a mainly Eurocentric society. Because the population more than tripled in less than a decade, many of the post-state newcomers were directed to relatively distant frontier settlements and new development towns in the Negev and Galilee regions (see Figure 1.1). The small and peripherally located development towns, established by the government to absorb new immigrants, often took on an ethnic character and came to be associated with low-income population, low-level economic development, and below-average social services. This dispersal created a core/periphery antagonism with a distinct spatial division of labour. The ethnic tensions inadvertently created at that time have remained a source of spatio-socio-economic problems.

Another significant aspect of the pre-statehood waves of immigration was the distinctly socialist ideology of the leadership of the newcomers. This, together with the hardships endured by these early European settlers, led to the creation of a variety of socialist or social democratic political parties, trade unions, and organisations which had a crucial influence on political and spatio-societal development. Thus, when the state came into being, the first governments were socialist-cum-social democratic coalitions, whose power bases extended throughout the societal infrastructure and had a major influence over international Jewish organisations. Not surprisingly, governmental control over economic affairs was widely felt, particularly since it had had a major say in the handling of considerable amounts of imported capital. This is not the place to discuss the ostensible paradox of a socialist regime allowing room for the development of a pluralist economy with a capitalist sector. Suffice it to say that over the years there has been a major shift in the government's role in changing the country's space economy, with increasing emphasis on private enterprise.

The current economy combines large government- and labour-owned economic sectors, including industries of communal settlements, with private enterprises ranging from small businesses to multinational corporations. The nature of Israel's industrial geography is therefore a reflection of this pluralistic economy.

Dimensions, physiography and climate

Another striking feature of the Israeli case is its small size. The country comprises 21,501 square km, with an additional 9,125 square km in the administered regions of Judea, Samaria and the Gaza Strip, occupied by Israel since the Six Day War in 1967 (see Figure 1.1). The spatial connectivity in the country is highly affected by its elongated shape. The road system of the north–south axis is about 550 km long, but only 120 km at its widest, narrowing (in the pre-1967 borders) to a mere 15 km in the central region.

Physiographically, Israel is divided into three north–south strips. In the west lies the Mediterranean coastal plain, widening gradually from north to south. Further east, a mountain range reaches elevations of up to 1,000 metres, interrupted by several east–west valleys. The eastern strip is part of the great Syrian–African Rift Valley; in Israel, a major part of this valley lies below sea-level, including the freshwater Sea of Galilee and the heavily salted water of the Dead Sea.

North to south, the climate changes from Mediterranean to arid. Despite the limited amount of cultivable land and the severe shortage of water for irrigation, Israel is self-sufficient in the production of a wide variety of fruits, vegetables, and flowers, as well as in poultry and dairy products – not only for direct consumption, but also to sustain a sizeable food processing industry and well-developed export markets. On the other hand, Israel imports grain, cereals, and fodder, as well as red meat and wood.

Population

The State of Israel is characterised by a dense and highly urbanised population that had reached some 5 million inhabitants by the end of 1990. Of these, 82 per cent were Jews and the rest mainly Moslem Arabs, Christian Arabs, and Druse.

From the beginning of the Jewish return to Zion, a dual economy – with a strong geographical dimension – developed, separating the Arab and Jewish populations. Growing hostilities between these two groups did not encourage confidence and trust, although practical and long-standing patterns of cooperation were established in several instances.

Since 1967, the addition of the Arab inhabitants of Judea, Samaria, and the Gaza Strip – by now approximately 1.7 million – has sharpened the dualism, and the space economy has been seriously influenced by this dichotomy. It has also had a major effect on the development of the labour market and on its spatial distribution.

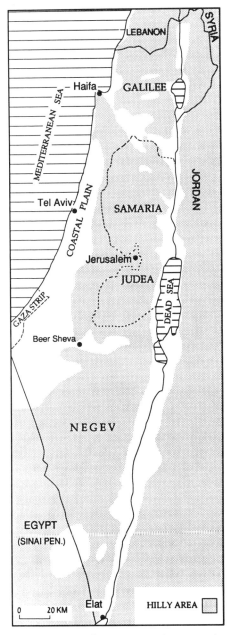

Figure 1.1 Israel – regions and topography

The percentage of the population in urban areas was about 80 per cent in the 1950s, rising gradually to more than 90 per cent in 1990, with major concentrations in the Tel Aviv, Jerusalem, and Haifa metropolitan areas.

THE PURPOSE AND STRUCTURE OF THE BOOK

The general aspects discussed above provide a major impetus to the contents and structure of this book. It is intended to serve those interested in Israel's geography and industrial development, as well as to provide a case study for those specialising in industrial geography and economic development. Since some international readers may have relatively little prior knowledge of Israeli society, geography, and economic development, we provide the requisite basic information. However, a thorough and innovate study relating the Israeli case to more general theories requires that we proceed beyond such basic descriptions and arguments that are familiar, or even self-evident, to Israel specialists. We therefore attempt to achieve a balance between offering basic description and analysis on the one hand, and presenting arguments and interpretations that we hope are new even to the Israeli expert on the other.

This book provides an overview of some of our own previous studies and attempts to integrate them with material gathered especially for the book. It should be noted that throughout the text we generally use the term 'industry' in the narrow sense of manufacturing and mining, although insights are provided into the location and development of linked economic activities as well. While industry, as we define it, comprises only some 20 to 25 per cent of the Israeli economy in terms of employment and product, it has a major role in determining its growth patterns. Moreover, geographically, industry has been the most flexible and mobile sector, and almost the sole agent in public efforts to reshape Israel's space economy. Hence, more than any other economic sector, it has influenced shifts in Israel's spatial structure.

We have chosen to present and examine Israel's industrial geography via an evolutionary approach, rather than by a regional survey. This decision was taken partly because of Israel's small size and partly because of the dominance of centrally controlled processes and policies in shaping the country's industrial geography. At the same time, we provide the reader with insights into the systematic and regional dimensions of Israel's industrial geography.

GENERAL BACKGROUND

The book emphasises several interrelated arguments concerning the role of structural, technological, and organisational transformations in industry; critical points in the Arab/Jewish conflict; waves of immigration; perceptions of development and planning; shifting economic and political realities; organisational dynamics; power relations within the Israeli political system; and the role of public initiative.

Chapter 2 provides a general background to Israeli industry. This is followed in Chapter 3 with the major arguments of the evolutionary study. We then provide a systematic overview of Israel's industrial geography from its pre-statehood roots to the post-Six Day War period (Chapters 4–7). Chapter 8 summarises certain changing realities in the 1970s and 1980s to which, we contend, the policy of spatial industrialisation failed to adjust. Chapters 9 to 15 examine the major issues in the country's industrial geography during these last two decades, including the slowdown and stagnation of the industrial base in development towns on the periphery; the decreased effectiveness of the government's incentive system due to inertia and incremental change; implications of externally controlled industry for the peripheral areas, as well as the crisis in the Federation of Labour enterprises, the major agents of industrialisation in the periphery; the rapid growth of high-technology industries and its spatial implications; the emergence of economic development strategies initiated at the local level; and the increasing interest in the local entrepreneurship option. Chapters 16 to 19 deal with the unique phenomenon of rural industrialisation, both in the cooperative Jewish, and in the Israeli Arab, sectors. Finally, Chapter 20 sums up the preceding discussions, with some insights into the 1990s, emphasising recent shifts in Arab/Jewish relations and also the recent mass immigration from the former USSR.

2

ISRAELI INDUSTRY
Background and role in the economy

GLOBAL SHIFTS AND ECONOMIC AND POLITICAL CHANGE IN ISRAEL

External economic and political conditions provide an essential background for understanding changing trends and spatial processes in Israeli industry. An overview of major shifts in this context in goals, objectives and means of Israel's spatial industrialisation policy is shown in Table 2.1. The divisions into stages of development are based in some cases on subjective judgement, and the table should be perceived only as a generalised framework for the study, rather than as an empirically tested outcome.

The period from the end of the Second World War to the late 1960s was one of rapid and steady global economic growth in the 'Fordist' mode of capital accumulation, characterised by mass-production, oligopolistic competition, and mass-consumption (Piore and Sabel 1984; Storper and Scott 1989). This period witnessed the rise of the Western welfare state, accompanied by productivity and wage increases and by an immense growth of multinational corporations operating under conditions of moderate international competition.

These years were also distinguished by rapid economic growth in Israel, interrupted only by a single, short, severe recession in 1966/7 (see Table 2.1). The Israeli economy was one of the fastest growing economies in the world during the two decades following the establishment of the state. Between 1950 and 1970, the gross domestic product rose annually by an average of 9.7 per cent – higher than the respective rates of growth in any country of a sample of twenty-five developed and developing countries, including Taiwan and Japan (Syrquin 1986). Despite the massive absorption of immigrants from war-ravaged Europe and poor Middle Eastern countries, the average annual per capita income rose during the same years by 5.2 per cent, a rate surpassed only by Japan, Greece and Taiwan.

The late 1940s and early 1950s were years of mass immigration, accompanied by shortages of basic needs that necessitated an austerity programme of rationing and price controls. Public investment was directed during these years towards the agricultural sector and housing construction. The accelerated development of import-substituting industry began in the mid-1950s, when investment capital became more readily available and the manufacturing sector was given high priority. Since 1962, a gradual shift from import-substituting to export industries has been evident (Halevi and Klinov-Malul 1968; Shaliv 1981; see also Table 2.1).

Somewhat isolated from global economic fluctuations during the British rule of 1918–48 (Nathan *et al.* 1946), Israel's integration with the international economy gradually grew with statehood, becoming increasingly dependent on foreign aid and the establishment of an export-oriented economic policy. Nevertheless, primarily as a consequence of the Arab boycott, foreign multinational investment in Israel was negligible, and most foreign capital consisted of various types of aid from foreign governments and world Jewry, as well as some direct investments by foreign Jewish businessmen. The 1967 Six Day War was a turning point in Israel's economic and political development. In addition to a change in the major goals of Israel's spatial policy (see Table 2.1), the French arms embargo and expanding defence demands motivated an immense expansion of military-related industries. Export-oriented industries and military equipment import-substituting industries were given priority during the fast recovery from the 1966/7 recession (Ben Porath 1986).

The late 1960s and early 1970s mark the transition of the world economy from a period of sustained growth to a period of instability and uncertainty. Two contradictory reactions can be identified: on the one hand, further internationalisation of production and finance and, on the other, disintegration of some vertically integrated industries. This latter process has led to a reversal of the long-term trend of decline in the role of small businesses (Storper and Scott 1989). What is less clear, however, is whether this disintegration is a temporary consequence of the global economic downturn (Amin and Robins 1990), or a more permanent reversal of the drive to

Table 2.1 General overview of external economic and political conditions, and shifts in the Israeli spatial policy

	External conditions		Spatial policy				
Decade	Global economy	Israel's political economy	Major spatial goals	Major goals of spatial industrialisation policy	Major objectives	Major means of implementation	Implementation
1950		1948 Establishment of Israel		construction and	Development of mineral extracting, government and agriculture-related plants in development regions	Ad-hoc decisions, direct investment individuals Labour Federation enterprises in development regions	Flexible, initiative of from above and below
		1948–1953 Mass-immigration, austerity, agricultural development					
	Rapid growth, stability, the height of the 'Fordist era'	1954–1965 Import substituting industrialisation, German reparations	Population dispersal	Job creation in the development towns	Promotion of industries insensitive to transport costs and independent of skilled labour in development regions	Grants, loans, tax exemptions and other incentives to plants located in development regions	Flexible, development from above
1960		1966–1967 Recession					
1970		1968–1973 The post-Six-Day War expansion based on import-substituting military equipment and exports	Population dispersal and settlement of the occupied territories				Formal organisational procedures, development from above

Year	Era	Period	Settlement	Goals	Industry	Local initiative	Political	Measures
1980	Slow growth, instability, restructuring processes. A 'post-Fordist era'?	1973–1977 The post-1973 war stagnation, transition from era of optimism to era of uncertainty / 1977–1984 The political upheaval, rapid inflation and economic stagnation, growth of high-technology industry	Settlement of the occupied territories	Job creation and reduction of regional inequalities	Diversification of manufacturing branch structure, dispersal of high-technology industries			
1985		1985–1989 Labour-Likud national unity govnt. lower inflation, stagnation, crisis in high-tech industry				Growing concern for the role of local initiative and entrepreneurship	Seeds of political initiative from below	
1990	Renewed growth?	1989–1992 Mass immigration from the USSR Likud-religious parties right-wing coalition	Settlement of the occupied territories and population dispersal	Job creation. Reduction of real distance between core and periphery?				As above plus additional measures to assist small business formation. Employment subsidies

'Fordist' mass-production methods. This latter process might be expected to lead to the resurgence of clusters of economic activity. Leading industries of the previous era have undergone a process of de-industrialisation, causing the loss of well-paid unionised employment in Western economies. The majority of new jobs created have been low-skilled and non-unionised (Bluestone and Harrison 1982). The weakening of welfare state mechanisms has been accompanied by structural change, most notably represented by growth of new high-technology industries (Hall 1985). It is too early to generalise, however, as to whether signs of renewed growth during the late 1980s indicate the onset of a new long wave of economic growth.

Israel has been particularly hard-hit during this period of transition in the world economy, shifting from a period of growth and optimism to a period of uncertainty. Apart from the world-wide economic shocks, a major repercussion of the 1973 Yom Kippur War was a very heavy economic burden of escalating defence and oil expenditures (Ben Porath 1986). From having one of the highest economic growth rates in the world during the 1950s and 1960s, Israel's expansion came to a virtual standstill during the 1970s and 1980s, lagging behind most Western nations. The annual growth of the business sector GDP went down from nearly 14 per cent, a figure it had attained from 1967 to 1972, to about 2.5 to 3.5 per cent from 1973 to the early 1980s (Bruno 1986). The entire business sector was squeezed by a large public sector and it has been suggested that Israel might have been premature in joining the developed countries in de-industrialising (Ben Porath 1986). An increase in inflation rates and an accumulating foreign debt were among the emerging problems during this period, while massive unemployment was avoided until the late 1980s.

In 1977, political upheaval was caused when the Labour party lost power to the right-wing Likud party (see Table 2.1). The new government liberalised foreign exchange, which proved to be disastrous (Bruno 1986), and pursued expansionary policies associated with new spatial goals of settlement in the occupied territories. A slight structural change represented by the rapid expansion of high-technology industries did not revive a process of economic growth. Only in 1985, under a joint Labour–Likud government, was inflation successfully controlled and the stabilisation of the economy achieved (see Table 2.1). However, the full burden of the prolonged economic stagnation was now being felt. Most of the 1980s, therefore, were characterised by increasing unemployment levels, static standards of living, the non-arrival of much-needed structural change, and unprecedented slow demographic and economic growth, accompanied by low rates of investment. The new wave of mass immigration from the former USSR, which commenced late in 1989, has been a major shock, altering previous trends. However, its long-term impact, either as a stimulus to new growth or as a source of great economic burden, is yet to be seen.

Shifts in Israel's industrial geography are assessed in this book in the light of these political and economic trends. Table 2.1 indicates that the goals and means of the spatial industrialisation policy laid out in the 1950s were relatively stable and responded to major shifts by small, gradual modifications. The following chapters treat in detail the various stages outlined in Table 2.1. However, some general attributes of Israeli industry must first be introduced.

NATIONAL COMPARATIVE ADVANTAGE

National comparative advantage in industrial development consists mainly of: (a) availability of natural resources; (b) access to capital; (c) low cost of other factors of production, such as labour and land; (d) abundant human capital; and, (e) geopolitical position. Of all of these properties, Israel enjoys a significant advantage only in human capital.

Availability of natural resources

Israel is not abundantly endowed with raw materials. Its major natural resources are non-metallic minerals, primarily potash and bromine from the Dead Sea and phosphate deposits in the Negev desert. The Dead Sea minerals are the only ones in which Israel enjoys a significant advantage over other producers, because these minerals are extracted through the use of huge evaporation ponds rather than by underground mining. Potash and bromine, used respectively as fertilisers and for a variety of industrial applications, are the main minerals extracted, but salts and magnesium chloride are also separated. Extraction of the Dead Sea minerals began in 1931, and since 1971 its commercial profitability has greatly increased. It is one of the major industrial sources of foreign currency in Israel.

Phosphate extraction began in the Negev in 1951, and became

profitable during the 1970s. However, the low market price per weight and the lack of any particular advantages of mining in the Negev made this enterprise more vulnerable to fluctuations of phosphate prices on the world market. Other non-metallic minerals extracted in the Negev – clay and quartz sand – are of marginal significance.

Limestone for the cement and building industries is also extracted in Israel. The only metallic mineral extracted commercially in modern Israel was copper from the Timna mines north of Elat. Production begun in 1959, but the mines were shut down in 1976 due to lack of profitability. Hence, the Dead Sea is the only natural resource in which Israel enjoys a particular advantage. Most other minerals are of low quality and require special treatment to be brought up to the quality acceptable in world markets. Industrial development in Israel based on natural resources has therefore been limited.

Energy sources

Israel also lacks significant energy sources. Small oil fields discovered in 1955 north-west of Beer Sheva supplied only a fraction of the country's needs and have long since been depleted. In the course of the many drillings which were made in order to discover new oil fields, small quantities of natural gas were found in several sites in the Negev. However, these were of little significance, supplying substantially less than 1 per cent of Israel's energy needs. Oil from the Sinai peninsula, conquered by Israel in the 1967 Six Day War, was of much greater significance and supplied 60 per cent of Israel's oil consumption during the mid-1970s. However, by 1980 the Sinai oil fields were all returned to Egypt as part of the peace agreement. Hydroelectric power potential is also negligible. Israel does enjoy a clear advantage in solar energy, but present technologies for utilising this energy source have limited its share to no more than 3 per cent of total energy needs. Hence, Israel has depended on imports for the supply of nearly all of its energy consumption. Israel's electric infrastructure is advanced and fully developed and is based on oil and coal-fired power stations; however, its geopolitical isolation inhibits any option of backup from the networks of neighbouring countries in the event of major failure or excess demand.

Access to capital and other factors

Israel does not offer easy access to capital for investment in manufacturing, nor to cheap labour or land. Furthermore, its geopolitical isolation in the Middle East and the Arab boycott make it more difficult for Israeli industry to penetrate large international markets. Only the availability of top-level qualified scientific and technical labour gives Israel a significant advantage, and this has facilitated rapid growth of high-technology industries since the late 1960s (see Chapter 13).

MAJOR ATTRIBUTES OF ISRAELI INDUSTRY AND ITS ROLE IN THE ECONOMY

Trends in the evolution of Israeli industry (see Figures 2.1–2.3) clearly reflect the major shifts in the Israeli economy described at the

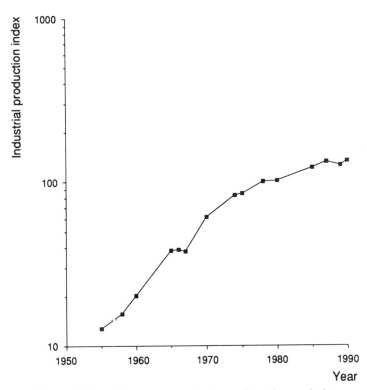

Figure 2.1 Israeli industry – production indices (not including diamonds), 1955–90 (1978 = 100)

Source: Central Bureau of Statistics, *Statistical Abstracts of Israel.*

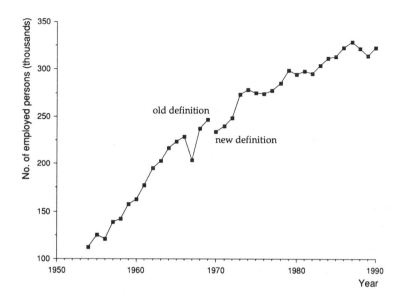

Figure 2.2 Israeli industry – number of employed persons, 1954–90

Source: Central Bureau of Statistics, *Labour Force Surveys*.
Note: Auto repair shops and repair services for households have not been classified with industry since 1970.

beginning of this chapter. The share of industry (manufacturing and mining) in the Israeli economy has been relatively stable since the 1950s, usually at somewhat less than one-quarter of the total labour force and net domestic product (see Figure 2.3). Hence, the growth in the share of tertiary activities in the Israeli economy has been mainly at the expense of primary activities. Between 1955 and 1965, industrial product tripled (see Figure 2.1), and the number of employed persons grew by 80 per cent (see Figure 2.2). The rapid growth of Israeli industry led to a slow rise in the proportion of industrial jobs of the total working population, with the exception of the recession years of 1966 and 1967. This proportion reached a high of 25.6 per cent of the total labour force in 1974 (see Figure 2.3).

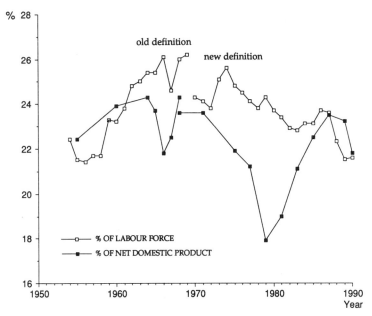

Figure 2.3 The share of industry in the Israeli labour force and net domestic product, 1954–90

Source: Central Bureau of Statistics.
Notes: Auto repair shops and repair services for households have not been classified with industry since 1970. Percentage of industry in NDP does not include the subsidy component in government loans to the industrial sector.

The post-1973 slow-down brought the increase in the number of employed persons in industry to a complete halt between 1974 and 1977. Hence, the share of industry in the labour force and the domestic product have declined. A declining rate of total growth in productivity since the 1970s has been attributed mainly to decreasing rates of capital utilisation and insufficient technological improvements, exacerbated by distortions in the government's policy of subsidising capital investment in manufacturing (Bregman 1986). Industry grew slowly and kept its share in the economy for most of the 1980s; however, during the years 1987–9, Israeli industry encountered a deep downturn. Employment and product decreased, and in 1989 the share of industry in the economy reached its lowest level since the late 1950s. This crisis was among the major factors for the increase in unemployment levels during

Year	Food	Textiles	Clothing	Wood products	Chemical products	Non-metal mineral products	Metal products and machinery	Electronic equipment	Transport equipment	Other[b]	Total
All plants – 1952 census classification											
1952	20.0	18.4		11.3	4.3	7.4	15.4	3.8	1.5	17.9	100
Plants with at least one salaried worker – old classification											
1955	18.2		18.3	8.2	5.7	9.0	15.0	3.1	3.3	19.2	100
1958	16.4		16.9	8.3	6.0	7.4	15.1	3.6	9.4	16.9	100
1961	15.5		17.9	7.7	5.4	6.8	15.6	4.2	10.6	16.3	100
Plants with at least five salaried workers – old classification											
1960	16.5		15.5	6.6	6.0	7.1	17.3	4.2	10.4	16.4	100
1965	15.3		18.8	7.1	4.6	6.1	16.4	4.6	11.7	15.4	100
1968	15.9		19.8	5.6	4.4	4.5	18.0	5.8	9.5	16.5	100
Plants with at least five salaried workers – new classification											
1965	16.2		19.8	6.9	4.8	6.5	18.1	4.8	6.6	16.3	100
1969	15.6	10.2		4.9	4.7	4.3	18.9	7.1	7.8	16.4	100
1972	14.8	10.3		4.9	4.5	4.4	19.0	8.7	8.1	15.2	100
Plants with at least one salaried worker – new classification											
1973	14.2	9.4	11.1	5.5	4.3	4.3	18.6	8.7	7.9	16.0	100
1977	13.9	7.0	11.0	5.1	6.2	3.7	20.1	9.2	8.1	15.7	100
1981	13.8	6.2	10.0	5.5	5.9	3.8	20.0	12.1	7.2	15.5	100
1985	14.3	4.9	10.1	4.5	6.2	2.9	19.8	14.9	6.5	15.9	100
1987	16.6	5.0	11.0	4.4	6.0	2.8	18.2	13.8	5.6	16.6	100
1990	15.1	4.1	11.4	4.7	6.0	3.2	18.0	14.2	5.7	17.6	100

Source: Central Bureau of Statistics, *Statistical Abstracts of Israel*.

Notes: [a] The Table does not include the diamonds branch. During the late 1960s, the classification of economic branches was changed. The definition of industry (manufacturing and mining) was narrowed, and auto repair shops and repair services for households were excluded. The major impact was in the transport equipment manufacturing branch, from which auto repair shops were excluded. The 1952 figures do not include slaughter houses, publishing, and auto repair shops. The 1955 figures do not include printing and publishing.
[b] Mining, leather, paper products, printing and publishing, rubber and plastic products, and miscellaneous.

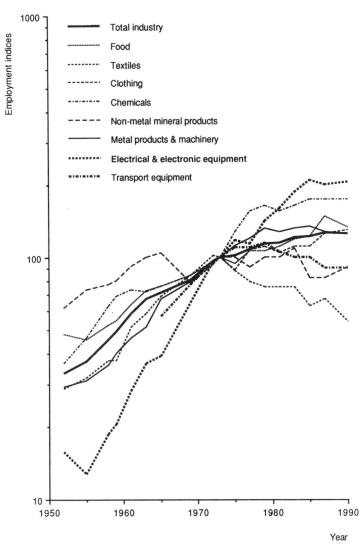

Figure 2.4 Israeli industry employment indices for selected branches, 1952–90 (1973 = 100)

those years. Industrial growth resumed in 1990 with the arrival of the new wave of mass-migration which fuelled local demand, especially for house-building materials. The industrial product grew

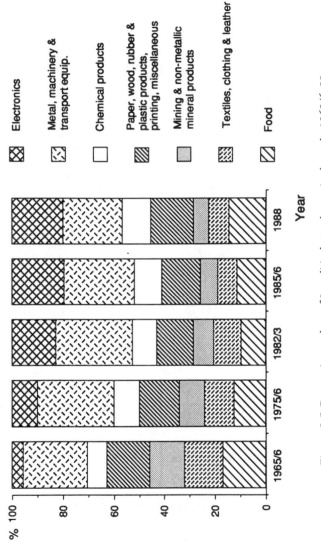

Figure 2.5 Domestic product of Israeli industry by major branch, 1965/6–88

Source: Central Bureau of Statistics, *Statistical Abstracts of Israel.*
Note: The figure refers to plants with five and more workers and does not include the diamond industry.

by 6.3 per cent and the number of employed persons in industry by 2.6 per cent. Modest growth continued in 1991, with a disruption during the Gulf War.

Periods of sluggish growth can be assumed to increase pressures for structural change. In Israel, however, it has been hard to prove the association between economic cycles and changes in the industrial structure. Structural change in Israeli industry was more rapid between the 1950s and 1970s, and slowed down during the later years of economic recession (see Table 2.2 and Figures 2.4 and 2.5). Foodstuffs, textiles, clothing, and metal products dominated Israeli industry during the early 1950s, and textiles and metal products in particular led industrial growth during the 1950s and 1960s. During the 1970s, chemicals, electronics, and metal products enjoyed the most rapid growth. Employment in the transport equipment branch (primarily Israel Aircraft Industries) grew rapidly during the post-Six Day War years, but has frozen since the 1970s. The slow industrial growth of the 1970s was primarily a consequence of stagnation in the more traditional low-skill branches, such as textiles, clothing, foodstuffs, and non-metallic mineral products.

A major problem of the 1980s was that hopes for a stimulating role of new leading industrial activities, particularly electronics, did not fully materialise. The number of employed persons in the chemicals and metal products branches failed to increase, and electronics, which had grown rapidly until 1985, suffered from recession in subsequent years. Most growth in industrial employment between 1983 and 1987 was in traditional industries such as foodstuffs and clothing. During the late 1980s, when these branches also began to contract, the recession encompassed most industrial branches. The crises in the military industries and in the Federation of Labour-owned enterprises, and renewed difficulties in textiles and clothing, were the major causes for industrial contraction during the years 1987–9. The mass immigration-driven growth of 1990 was most evident in traditional branches such as non-metallic mineral products, wood products, textiles, clothing, and basic metals. These branches benefited most from the expanding demand of the construction industry and the final consumption demand of the immigrants themselves.

Industrial exports grew rapidly during the early years of the state. Agricultural exports were still dominant in 1949. Moreover, while low value-added diamonds comprised exactly one half of the industrial exports of that year, most of the remainder were citrus fruit industrial products (see Table 2.3). By 1959, however, industrial

Table 2.3 Israeli industrial exports, 1949–90 (in millions of US dollars)

Year	Total industrial exports	Industrial exports (not including diamonds)	Industrial exports (including diamonds) of total goods exported (%)
1949	10.4	5.2	36.5
1950	18.1	9.3	51.6
1955	54.9	34.7	61.6
1960	153.5	92.6	70.9
1965	343.1	189.4	79.9
1970	637.5	392.9	81.9
1975	1,610.6	969.9	83.0
1980	4,955.5	3,340.4	89.5
1985	5,667.6	4,234.9	90.5
1988	9,131.6	6,294.5	93.6
1989	10,113.9	7,034.1	94.8
1990	11,060.0	7,823.9	95.5

Source: Central Bureau of Statistics, *Statistical Abstracts of Israel*.

exports, excluding diamonds, exceeded agricultural exports, and by the 1980s, industrial goods were more than 90 per cent of Israel's exports. The composition of industrial exports changed remarkably during the last two decades (see Figure 2.6). In 1970, textiles was still the leading industrial export, followed by foodstuffs, and these two traditional branches were responsible for more than 40 per cent of the total exports in this sector, excluding diamonds. In 1980, chemicals and metal products were already leading. By 1989, the electronics sector, which had previously relied mainly on local defence contracts, was a major exporter, with over 20 per cent of the total industrial exports. The share of textiles and foodstuffs together was now down to only 17 per cent (see Figure 2.6).

AN INTERNATIONAL COMPARATIVE PERSPECTIVE

The Israeli economy has been classified among the lower-income group of developed market economies, or among the high mid-income nations (UNIDO 1983; World Bank 1988). On the basis of theoretical constructs relating industrialisation and development (Chenery *et al.* 1986), Bar-El *et al.* (1989) showed that the developmental stage of Israel has been well above that of the developing

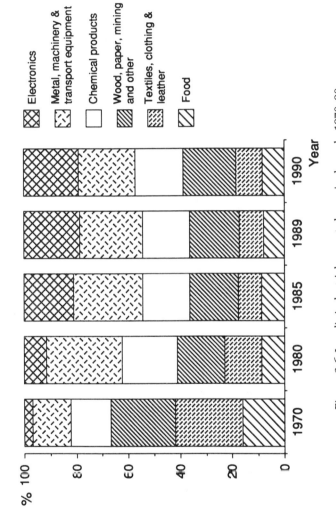

Figure 2.6 Israel's industrial exports by major branch, 1970–90
Source: Central Bureau of Statistics, *Statistical Abstracts of Israel*.
Note: The figure does not include the diamond industry.

countries; however, it has still been below that of the industrialised nations. Based on data derived from 1982 input–output tables, Israel would rank forty-first among the non-centrally planned economies in terms of value added in manufacturing, and twenty-fifth among the non-centrally planned economies in terms of value of industrial exports (Dicken 1986). According to UNIDO (1985), Israel ranked seventh in terms of the R&D orientation of its export goods sector among thirty-two countries, eighth in terms of R&D expenditure per worker, and fifteenth in terms of the R&D orientation of imported products.

The share of industry in the labour force in Israel does not differ substantially from that in Western industrialised countries (Ministry of Industry and Trade 1987). Industry had a greater share in industrialised Western economies until the 1970s, but the subsequent decline in the proportion of industrial employment and output in the economy of these countries has led to the closing of this gap. However, productivity in Israeli industry remained much lower than in Western industrialised countries, and the capital investment per worker was relatively low in most branches. These differences were reflected by wages in Israeli industry, which in 1985 were about one-third of those in the United States – much lower than wages in Canada and Western Europe but higher than wages in countries such as Greece, Portugal, Hong Kong, and Korea (Ministry of Industry and Trade 1990).

The current state of Israeli manufacturing industries has been characterised as being in a transitional stage from quantitative growth to a qualitative change (Bar-El *et al.* 1989). This change is associated with increasing international linkages, a declining proportion of output accounted for by traditional industries, and a comparative advantage in those industries using capital-intensive means of production and highly skilled labour. This process leads to an increasing diversification of the industrial base. Data for the early 1980s show that 47 per cent of the Israeli labour force were employed in branches defined as 'growing' as opposed to 22 per cent in 'traditional' branches, whereas the figures among the five most industrialised market economies were 55 per cent and 15 per cent, respectively (Bar-El *et al.* 1989). Nevertheless, the prolonged stagnation of the Israeli economy has been associated with a slow-down in the progress of Israel's industry through the transitional stage and its lagging behind in the pace of development relative to rapidly expanding East Asian and Mediterranean economies.

Part II

THE EVOLUTION OF THE INDUSTRIAL GEOGRAPHY OF ISRAEL UNTIL 1973

3

INDUSTRIAL EVOLUTION IN ISRAEL
General arguments

The industrial geography of Israel at the outset of the 1990s is a product of spatial processes that were initiated during the period of British rule, and of spatial policies implemented by the government to modify these processes thereafter. The foundations of modern industry in Israel were mostly laid down by Jewish immigrants during the 1920s and 1930s, and consolidated during the Second World War. These years witnessed the evolution of a new Jewish urban system which has basically endured and consolidated as the core of present-day Israeli urban structure. Since the establishment of the State of Israel, persistent efforts to modify this spatial structure, which has been constantly reinforced by spatial economic processes, have been made. These efforts have included the establishment of dozens of new non-metropolitan development towns and the adoption of an industrial dispersal policy. Therefore, in order to evaluate present issues and problems in the industrial geography of Israel, one should focus on the implementation of Israeli spatial industrialisation policy, which has been remarkably consistent for nearly forty years. This policy has served as a major means for population dispersion and has attempted to provide solutions to economic problems of Israel's development towns. A comprehensive assessment of the spatial industrialisation policy should consider the interaction of economic shifts, political situations, changing planning perceptions, organisational dynamics, and the actions of pressure groups and entrepreneurs. Such an assessment, which begins with circumstances prevailing at the time when the policy was formulated, is indispensable for a correct evaluation of spatial trends and shortcomings of the policy during the post-1973 years of economic stagnation (see Part III).

The historical overview presented in the following chapters

emphasises several major interrelated arguments concerning the spatial industrialisation policy and the evolution of Israel's industrial geography:

1 Structural, technological and organisational transformations in industry have influenced Israel's industrial geography, since new industrial activities have different spatial behaviour than that of older activities. Interpretation of the spatial consequences of these transformations has relied substantially on industrial location theory, stressing the role of concepts such as product life cycle, and the role of transaction costs in the formation and decline of industrial clusters in various manufacturing activities such as diamonds and electronics.
2 Critical events in the Arab/Jewish conflict, as well as waves of immigration, had a crucial influence on Israel's human geography, including its industrial structure. These unique factors led to a particular form of economic development and spatial structure that has deviated from generic global trends. However, the isolation and uniqueness of the Israeli case has gradually diminished over the last decades, as integration with, and dependence on, the world economy has grown.
3 Theories of development and planning perceptions have diffused slowly from leading countries in these fields to local Israeli planners, and from them to Israel's decision-makers. Hence, proposals for reorienting spatial industrialisation policy were generally based on realities of previous periods. As a result, planning programmes were frequently outdated by the time they were implemented. This was the case with early spatial strategies that emphasised ideas based on classical industrial location models, as well as those strategies based on ideas resem- bling Christaller's central-place theory, which suited conditions of the nineteenth and early twentieth centuries. Strategies based on the British Barlow Report (Hall 1982), for instance, were implemented in Israel seventeen years after the report was published and twelve years after its implementation in Britain. Later economic development strategies emphasising diversification and various forms of localised initiative and entrepreneurship have also been adopted with a significant time lag.
4 The effectiveness of Israel's spatial industrialisation policy has declined, due to its failure to adjust to the new economic and political realities of the 1970s and 1980s. According to Ben Porath

(1986), the ideology and institutions that were forged in Israel during the 1950s and 1960s based on large numbers of immigrants and rapid growth were ill-suited to cope with the challenges of the 1970s and 1980s. This caused misallocation of resources, prevention of needed structural change in the economy, and the increase of foreign dependence. No major intentional reorientation of spatial industrialisation policy has been made since the 1950s despite major changes in the economic and political external conditions after the 1967 and 1973 wars. These changes included: (a) the growing role of exporting industries and military import-substituting industries; (b) the penetration of Arab labour from the occupied territories into the Israeli labour market; (c) new spatial-political goals; (d) reduced economic growth rates; and, (e) shifts in global economic conditions. These changing conditions were disregarded for a long time. Assessment of the causes for the limited impact of the policies implemented during the 1970s and early 1980s emphasised changing internal socio-economic conditions in the development towns. Hence, proposed modifications were only incremental and were based on perceptions of rapid economic growth, which had not been the case since 1973.

5 The evolution of spatial industrialisation policy has been influenced by organisational dynamics, which characterises many public and private organisations. According to theories relating growth and structure of business organisations, young business organisations are usually managed in an informal entrepreneurial manner. The transformation into centralised functional management is required at some point of growth and maturity, and further transformations into more decentralised structures at later stages (Penrose 1959; Greiner 1972). Stable periods of growth generally end with management crises which can either lead to the dissolution of the organisation or to the adoption of new forms of management. Organisational decline can result from both external and internal pressures (Adizes 1988). A typical tendency to cling to measures and procedures which have proved to be efficient in the past, despite marked changes in external conditions and pressures, is one source of organisational decay. However, inability to adjust to internal pressures as the organisation matures can bring about similar symptoms. These symptoms include excessive conservatism and inertia, apathy and alienation of workers, and reduced effectiveness, efficiency,

and, in the case of business organisations, profitability.

Industrial dispersal in Israel was initially carried out through *ad hoc* decisions, leaving room for the initiative of individuals at the national and local levels. Later, a consistent policy was put into effect through the initiative of Pinhas Sapir, who was then Minister of Commerce and Industry. Following a logical sequence of organisational development, the flexible entrepreneurial activity of leading personalities was gradually replaced by formalised bureaucratic procedures. However, these established procedures were among the main factors delaying necessary adjustments. Delays in adjusting business organisations to new dimensions and to changes in the environment generally emerge in the form of financial losses. However, in the case of public organisations and policies, need for adjustments can be disregarded for years, despite clear symptoms of organisational decay, as failures are less measurable and barriers to change are greater. A shift to a more decentralised spatial industrialisation policy may recently have begun after a long period of stagnation.

6 Power relations within the Israeli political system have been of crucial importance in the evolution of Israel's spatial industrialisation policy. Changing government relations with the Federation of Labour enterprises and with the private sector were a major factor influencing early formulation of this policy, and shifting balances of power between central government and local authorities have greatly influenced the present performance of the policy.

7 The decisive role of 'public' entrepreneurs, or power brokers, in shaping and reshaping industrial dispersal has been prominent under certain conditions. The ability of a public official or a public agency to shape new processes of change has been extensively debated in the past. A classical example in the field of spatial planning is Robert Moses, a public entrepreneur viewed as a vital force in the shaping of New York for several decades and as a person who influenced the development of American cities more than any other person in the twentieth century (Caro 1975). An alternative view maintains that determining major urban processes of change is beyond the control of the 'institutional' entrepreneur, or any other small group of actors (Danielson and Doig 1982). In any case, identifying periods when such personalities tend to proliferate, and studying their views and styles of action, is important for the evaluation of the policy during these periods.

Based on these generalisations, two major qualitative scales are utilised for the schematic description of stages in the development of the Israeli spatial industrialisation policy (see Figure 3.1). The first scale distinguishes between phases of development 'from above', in which central government and other nation-wide organisations are dominant, and phases of development 'from below', in which local government and local groups and entrepreneurs are more active. The second scale distinguishes phases of flexible activity of public agencies from phases dominated by more formal criteria and established organisational procedures.

The major attributes of each stage defined in Figure 3.1 are summarised in Table 3.1, which also refers to an initial pre-statehood stage and to a post-1990 transformation associated with mass immigration. Major clusters of modern industry began to evolve in the coastal plain during the pre-statehood years of *laissez-faire* economic policy. Efforts to disperse industry in Israel began through uncoordinated initiatives of individuals at both the local and national political levels. Only at a second stage were principles and organisational tools formulated through the strong initiative of a key personality in the central government. The third stage marked a transition from informal entrepreneurial policy to more formalised

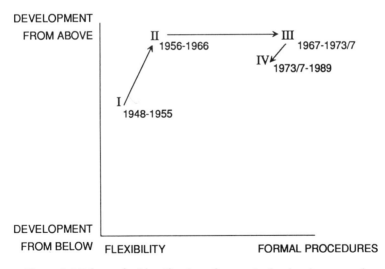

Figure 3.1 Scheme for identification of stages in the development of Israel's spatial industrialisation policy

Table 3.1 A summary of stages in the development of Israel's spatial industrialisation policy

---

Stage 0 – Pre-1948
1. *Laissez-faire* policy of the British government. Lack of intervention of Zionist organizations in industrial development.
2. Evolution of major modern industrial clusters in the Tel Aviv and Haifa areas, and stagnation of traditional centres of crafts in mountainous towns.

Stage I – 1948–55
1. Flexible policy. *Ad hoc* decisions and no formal procedures – initiative of individuals at the national and local political levels.
2. Limited planning experience and resources.
3. Pioneering industrial investment of government and Federation of Labour enterprise in development towns.
4. Dispersal of raw material-oriented or market-oriented plants.

Stage II – 1956–66
1. Flexible policy. Development from above – the dominant role of initiative of a key personality in the central government.
2. Formation of a system of capital incentives according to the British planning tradition; definition of development zones. Formulation of organisational tools.
3. Externally-owned private industry penetrates the development towns with massive government support.
4. Rapid industrialisation of the periphery. Major criteria: footloose industries, offering ample employment for workers with no previous skills. Dispersal of capital-intensive plants employing unskilled labour.

Stage III – 1967–73/7
1. Formal organisational procedures. Development from above – institutionalisation of the policy through statutory measure.
2. Increasing difficulties in adjusting the policy to external and internal change and to emerging competing goals.
3. Growth, but non-renewal, of the industrial base in development regions.
4. Local supply of jobs in development towns is dominated by non-growing branches and becomes less suitable to changing local demand.

Stage IV – 1973–89
1. Reduced effectiveness of policy from above, due to changing realities and diminishing public resources. Seeds of political initiative from below from local authorities.
2. Political interests and bureaucratic inertia hamper more than incremental changes in existing government policies.
3. Improving conditions in some favourably located development towns enjoying active political leadership. Crises in remote towns.
4. Calls for diversification of the industrial base in the periphery. Emerging interest in local entrepreneurship.

Stage V – 1990–?
1. The shock of mass immigration. Reinforcement and modification of policies from above. Unclear consequences as to the balance of power between central government and local authorities.
2. Implementation of new measures to assist entrepreneurship.
3. Calls for shifting priorities from direct capital incentives to industry to concentration of investment in the transportation and communications infrastructure.

bureaucratic procedures. This transition, while securing long-term continuation of the policy, made adjustments to new political and economic realities increasingly difficult. Thus, in the fourth stage, the effectiveness of the policy may have been further reduced, thereby encouraging the emergence of local initiative and action from below.

The general arguments presented above and their consequences for the evolution of the spatial industrialisation policy apply also to other countries that have spatial policies similar to those of Israel. Surprising similarities can be found in the evolution of spatial industrialisation policies across countries, due to similar external economic pressures, planning perceptions and organisational dynamics. The Italian regional policy (Martinelli 1985; Tommel 1987) presents an interesting case for comparison. Regional policy crystallised at about the same period in both countries, and after an initial period of focusing on agriculture, emphasis shifted to manufacturing. The capital incentives system was successful in dispersing heavy industry and other mass-producing externally-owned plants. Nevertheless, similarly to the Israeli case, inertia and established structure of interests hindered reforms in the policy when new realities of the international division of labour and European integration began to emerge. Growth opportunities for Italian industry have shifted more and more towards flexibly operating, small-sized enterprises, and classical instruments of regional policy have lost their stimulating effect (Tommel 1987). As in the Israeli case, it has been suggested in Italy that capital subsidies should be made more accessible to smaller and less-organised firms, and a system of technical and organisational assistance should be developed (Martinelli 1985). An international comparative perspective can provide, therefore, insights both into the evolution of spatial industrialisation policies elsewhere and into generalities and specificities of the Israeli case.

4
THE PRE-STATEHOOD ROOTS

AN OVERVIEW

Although seeds of modern industry can be traced to the late nineteenth century, the Land of Israel, which was then Palestine, remained basically a traditional pre-industrial region throughout the era of Turkish-Ottoman rule until its termination in 1918. The industrialisation process did not really commence until the period of the British Mandate (between 1918 and 1948). However, the industrial sector remained rather small in absolute terms and its growth was limited due to the non-protectionist *laissez-faire* policy of the British government (Nathan *et al.* 1946; Shaliv 1981). Zionist organisations were engaged mostly with agricultural settlement, and industrial development was left to the rather weak private sector (Gaathon 1963).

Industrialisation was influenced by several major forces during the period of the British Mandate. These forces included:

1 *Waves of Jewish immigrants*: The arrival of immigrants in large numbers stimulated growth, particularly through the infusion of small investment capital and increased local demand. Construction has always become the leading sector during waves of new immigrants, and it stimulated the rest of the economy (including manufacturing industries) to expand. The foundations of modern industry in Israel today were laid down in the Palestine of the 1920s and 1930s, primarily by Jewish immigrants from Central European countries who arrived with some capital, entrepreneurial skills, and experience in manufacturing.
2 *Deteriorating Jewish–Arab relations*: Industry in Palestine evolved under conditions of two populations coexisting in a state of conflict. The deteriorating relations between Jews and Arabs

through a series of clashes during the years 1920/1, 1929 and 1936–9 enhanced the evolution of a dual economy in which initial ties between the rapidly growing Jewish sector and the slower-growing Arab sector were frequently disrupted. The two sectors differed greatly with respect to the character of industries: their types, the size of industrial plants, capital invested per worker, and locational patterns. By the final years of British Mandate rule, two nearly separate economic systems, Jewish and Arab, were operating in the same territory. Thus, aggregate measures of industrial growth and spatial organisation, disregarding the dual nature of the economy, obscure real processes.

3 *The Second World War*: This led to a great expansion of industry in Palestine. Mediterranean shipping routes were blocked, forming a 'natural' protective barrier for local industry, and demand generated by allied forces stationed in the Middle East fuelled local production (Nathan *et al.* 1946). Whereas construction was the leading sector in periods of growth stimulated by immigration, military-supplies-manufacturing industries led wartime growth. As a result, local industry underwent a significant restructuring.

The evolution of modern industry created a major shift in Palestine's industrial geography. This shift was partly a consequence of locational and political factors which led Jewish settlements to concentrate in the coastal plain and the northern valleys, since the mountainous areas were already densely populated by Arabs. Location requirements of modern industries, which differed from those of traditional crafts, also contributed to a modification of past spatial patterns. Accordingly, Jewish-owned industry showed a high tendency to be concentrated in the coastal plain, especially in Tel Aviv and Haifa, whereas Arab industry remained spatially dispersed in coastal and mountain towns alike. As a result, Jerusalem and other mountainous towns lost their dominance in favour of the central coastal plain, with Tel Aviv and Haifa becoming the centres of light and heavy industry, respectively. These two coastal cities competed for a prominent role as industrial centres, but it was Tel Aviv which evolved into the main concentration of population and industrial activity because of its more favourable location with respect to local markets and links within Palestine. The major economic attributes of these two largest industrial centres in Israel, which persist to the present day, were thus formed during those pre-statehood years and set the stage for early spatial concerns and policies adopted by the government of Israel after 1948.

END OF THE OTTOMAN ERA – TRADITIONAL LOCAL SPECIALISATIONS AND SEEDS OF MODERN INDUSTRY

Manufacturing capacity in Palestine at the end of the nineteenth century consisted mainly of primitive, low investment, manually driven, small plants. It included a food processing sector dotted by water-powered flour mills, olive-oil presses, and tobacco and dry fruit processing, as well as small workshops producing pottery, leather products, wool clothing, and simple metal goods. The many carpenters, shoemakers, tailors, locksmiths, and building materials suppliers were also counted among the people employed in manufacturing. Most urban centres supplied these products for their local markets and nearby hinterlands (Ben-Arieh 1981). Nevertheless, certain towns specialised in particular commodities (see Figure 4.1). The largest city at that time, Jerusalem, enjoyed an influx of immigrants, including some foreign interests during the second half of the nineteenth century; consequently, the related massive construction industry promoted the development of tile manufacturing. Another flourishing craft in Jerusalem was the printing and bookbinding business, reflecting the city's character as a religious centre and seat of learning.

Among the other land-locked mountain towns, Nablus (Shekhem) had the most prominent industrial specialisation. Between fifteen and twenty enterprises in Nablus were engaged in olive oil and soap production, which were among the earliest of Palestine's goods exported throughout the Middle East. Hebron specialised in blown glass products, Bethlehem in souvenirs sold to pilgrims, and Nazareth in knife and sickle products. Among the coastal plain towns, Gaza was a major centre, specialising in linen and rug manufacturing. In other towns on the coastal plain, manufacturing was less prominent.

Hence, the industrial geography of Palestine was a typical example of pre-industrial local craft specialisations based on traditional skills and availability of particular raw materials or markets. The Ottoman rule did not contribute to the promotion of industrialisation, and corrupt administration and myriad bureaucratic and legal problems confronted entrepreneurs and developers. None the less, the pace of change became more marked after the 1880s. While

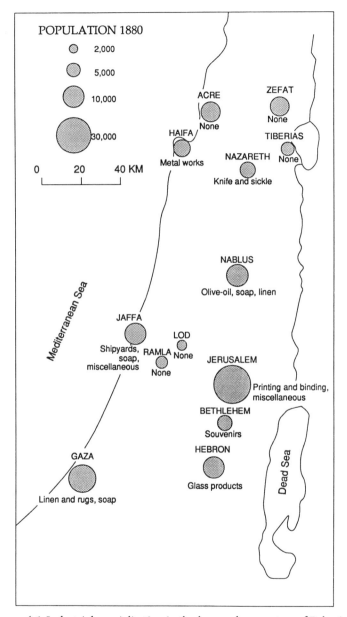

Figure 4.1 Industrial specialisation in the large urban centres of Palestine during the late Ottoman period

Source: After Ben Arieh (1981).

exposure of the economy to imports set in motion a long process of decline in local traditional crafts, such as clothing, early attempts were made to establish modern factories by non-Jewish and Jewish immigrants from Europe. These attempts were concentrated partly in Haifa, but mostly in the Jaffa area which served as the port of entry to Palestine and as the new centre of foreign trade.

The German Templers, who began arriving in 1868, played a major role in the introduction of modern industry and crafts in Palestine. They became engaged in metalwork, horse-driven carriages, building materials and other handicrafts in their Haifa and Jaffa colonies (Carmel 1973). In 1895, a Templer family called Wagner established a metalwork factory in Jaffa which produced pumps, engines and other machinery. The Wagner brothers' factory, considered to be the largest and most advanced in pre-First World War Palestine, employed 150 workers at its peak in 1908.

Jewish immigrants motivated by Zionist ideology arriving after the 1880s sowed the seeds of modern Jewish industry primarily in Jaffa, and later in Tel Aviv (Beilin 1987). A pioneering effort was the metalwork factory founded in 1887 by Leon Stein, a Jewish immigrant of Polish origin. At its height in 1909, this factory, which was Wagner's major competitor, employed 150 workers before collapsing a year later. Two modern oil factories were established in 1906 near Lod and in Haifa. Both failed shortly afterwards. A large plant, which is still in existence today, was the Carmel wineries, established by the Baron Edmond de Rothschild during the 1880s in an attempt to enhance the economic base of the colonisation efforts at Rishon LeZiyyon (Avitsur 1976). In the last years of Ottoman rule, Jews were dominant in modern industrial development, and while many ventures failed, they set the basis for further industrialisation under the governance of the British Mandate.

THE BRITISH MANDATE ERA – INDUSTRIAL DEVELOPMENT IN A DUAL JEWISH–ARAB ECONOMY

The switch from the Ottoman Empire to British rule was associated with substantial economic change. Considerable industrialisation took place during the early 1920s as Palestine recovered from the severe crisis caused by the First World War (Binah 1924). New enterprises were generally still in traditional industries geared to local demand, and the main departure from the past was in terms of the size of the new plants and the equipment they utilised. The tendency to move from primitive, small-scale manufacturing to large-sized plants was not only due to the more efficient British administration but also to the territorial extent of the British occupation and improved transportation and communications.

Since the 1920s, industrial development has demonstrated a clear causal relationship between immigration and economic growth as represented by the rate of increase in capital stock (Ben Porath 1986). The arrival of the Fourth Aliyah (immigration wave) from central Europe between 1924 and 1926 led to the increase of the number of industrial plants in the Jewish sector from 279 to 547 in less than two years (Beilin 1987). The crash of 1926/7, caused by the inability to continue financing rapid economic expansion, was as severe as the preceding boom. A report issued by the Zionist movement counted no less than 101 plant closures in 1926.

The results of the first comprehensive census of industries taken in Palestine (Palestine, Department of Customs, Excise and Trade 1929), analysed by Eliachar (1979) and Gross (1979), shed light on those early years of the industrialisation and evolution of the dual Jewish–Arab economy. Foodstuffs and chemicals dominated, and together they constituted about 40 per cent of the employed persons in industry in 1928 (see Table 4.1). It should be noted, however, that the chemicals sector was then mostly composed of oil-pressing and soap manufacturing. In 1928, Jewish-owned firms comprised 44 per cent of Palestinian industrial employment and product, and 64 per cent of the fixed industrial capital (Gross 1979). The Arab industrial sector was still larger than the Jewish one in terms of employment, but the average number of employed persons per factory was higher in the Jewish sector, and the average investment in a Jewish-owned factory was over four times larger. Nearly one-third of Arab industry was engaged in olive-oil and soap production, whereas Jewish industrialists were more heavily represented in clothing, printing, wood products, and construction supplies. Specialisation in clothing and printing reflected cultural differences between the Jewish and Arab populations, while wood and construction supplies were indicative of the construction boom that accompanied the waves of Jewish immigration.

Contrary to the world-wide recession of the 1930s, the economy of Palestine embarked at this decade on unprecedented economic and industrial growth (see Table 4.2). This growth was inspired by waves of Jewish immigrants fleeing Europe. In the years 1933–5, a new wave of tens of thousands of immigrants, known as the Fifth

Aliyah, came pouring into the country. With them came considerable amounts of capital and industrial experience. Among the newcomers were many former industrialists from central Europe, who brought with them not only their technical experience but also modern industrial equipment which they were permitted to take along in return for the capital they had to leave behind. Hence, the mass influx of these immigrants and the accompanying prosperity made industry a primary factor in Palestine's economy.

The years 1936–9 were marked by an economic slow-down which paralleled the Arab uprising and civilian disturbances. Restrictions imposed by the British on Jewish immigration further

Table 4.1 Palestine – employed persons in industry and size of average plant by population group and industrial branch, 1928[a]

Industrial branch	Employed persons [b]			Average no. of employed persons per plant		
	Jewish	Arab	Total	Jewish	Arab	Total
Food and tobacco	20.7	20.5	20.6	10	7	8
Chemicals	4.3	31.7	19.7	18	6	6
Wood products	9.4	6.1	7.5	4	3	3
Paper and printing	10.3	1.8	5.5	10	9	10
Linen	7.6	8.9	8.3	21	3	4
Clothing and toiletries	19.0	9.6	13.7	4	2	3
Leather and canvas	1.2	1.4	1.3	5	3	4
Quarries and salt	3.9	5.5	4.8	51	4	6
Bricks, cement, etc.	9.9	5.7	7.5	20	5	9
Metalworks	9.5	7.3	8.3	6	4	5
Jewellery	0.8	0.8	0.8	2	2	2
Other industries	1.3	0.7	0.9	6	3	5
Electric power	2.3	0.1	1.0	30	2	19
Total (absolute no.)	7,869	10,086	17,955	7	4	5

Source: Eliachar 1979.

Notes: [a] Metzer and Kaplan (1990) point out several inaccuracies in the table – amongst them: the over-enumeration of workers in the chemicals industry despite their being employed only one month a year, and the omission of craftsmen producing on barter rather than on a market basis. Hence, they estimate the correct figures for the total number of employed persons in industry was 10,661 in the Jewish sector, and 16,136 in the Arab Sector
[b] Percentage of employed persons in industry out of total population group.

Table 4.2 Palestine – employed persons in the Jewish industrial sector, 1921–43

Year	Number of employed persons	
	Industry	Handicrafts
1921/2	4,750	
1925	4,894	n.a.
1930	7,582	3,386
1933	14,419	5,176
1937	21,964	8,022
1943	45,049	n.a.

Source: Gurevich et al. 1947.

acted to limit local demand. The turning-point came during the Second World War, particularly after 1941. Growing transport difficulties in the Mediterranean and increasing demands of allied forces stationed in the Middle East generated conditions of full employment and utilisation of other production factors. The number of industrial labourers doubled, and most enterprises worked multiple shifts. New industries were introduced, such as precision tools, industrial and agricultural machinery, ammunition, and chemicals. A whole new diamond-polishing industry was developed by Jews fleeing Belgium and the Netherlands following the Nazi occupation (Gurevich et al. 1947). By 1943, product of the Jewish industrial sector surpassed that of Jewish agriculture. The end of the war marked the onset of a transition period in industrial development, which paralleled increasing tensions between Jews and Arabs, and which reached its climax in the 1948 War of Independence.

Agents of industrialisation during the British Mandate

Four agents of industrialisation played roles of varying significance during the period of British Mandate: prominent immigrant entrepreneurs, Jewish organisations, the British Mandate government, and Arab investors. None the less, until 1948 industrial development in Palestine was mainly the result of the efforts of entrepreneurial individuals, rather than those of government or other public agencies (Avitsur 1985; Liwschitz-Garik 1946).

First and foremost among the privately initiated large firms was the Palestine Electric Corporation (PEC) founded in 1923 by Pinhas Rutenberg, an immigrant from Russia. Although privately owned electric generators were introduced to Palestine as early as 1909, the earliest electric power plants providing electricity on a commercial basis were constructed in Tel Aviv, Haifa and Tiberias during the early 1920s. Thus, the industrial census of 1928 found 404 plants out of 3,505 that were powered by electricity. In 1926, Rutenberg, via the PEC, received a concession to supply electricity throughout Palestine, except for Jerusalem. A hydroelectric power plant was then constructed at the conjunction of the Jordan and Yarmuk Rivers south of the Sea of Galilee. Inaugurated in 1933, this sole hydroelectric power plant ceased operation in 1948, when the section of the Jordan River that it utilised became the cease-fire line between Israel and Jordan.

Another major industrial activity, which also started as a concession issued by the British Mandate in 1929, was potash production from the Dead Sea initiated by another immigrant from Russia, Moshe Novomeysky. Operations began in the accessible northern section of the Dead Sea in 1931 and expanded to the shallower waters of the southern Dead Sea in 1937 (Novomeysky 1946). Both enterprises were essentially nationalised after the establishment of the State of Israel.

Zionist organisations were leading agents for development in Palestine. However, they were concerned to a large extent with promoting rural-cooperative settlements. This was partly due to the 'return to the land' ideology, itself a reaction to the occupational composition of Jews in the Diaspora. It was also due to anti-urban socialist ideologies which had dominated the Zionist movement since the early twentieth century (Cohen 1970). As Jewish–Arab relations deteriorated, and proposals to divide Palestine between the Jews and the Arabs emerged, agricultural settlement became increasingly important as the most effective method of securing large tracts of land to Jewish hands. Hence, little attention or support was given to the industrial sector by Zionist organisations (Beilin 1987).

The pre-state Jewish organisation that was actively involved in industrial development was the General Federation of Labour (the Histadrut). Its ideology approved of developing enterprises owned by the working class through their representatives. After an early emphasis on construction, the Federation of Labour became increasingly involved with industrial development during the 1940s. Its construction concern, Solel Boneh, established Koor Industries in 1944 as a subsidiary for developing and managing industrial plants (Biltski 1974).

The British Mandate government was not involved directly in industrial investment, except for the oil refineries and transportation-related repair shops. However, its efforts in modernising the transportation and communications infrastructure and in maintaining a stable and reliable public administration were of great significance for industrial development. The British government refrained from highly visible protectionist measures, largely on the grounds that such a protectionist industrial policy would discriminate against the Arabs; its beneficiaries would be Jewish-owned modern industries, whereas Arab consumers would bear the cost of higher prices. The Arabs fiercely opposed any public effort to promote industrialisation in the Jewish sector, since such efforts were perceived to increase Palestine's carrying capacity for additional immigrants.

Substantial industrial development also took place in the Arab sector, but at a slower pace. The industrial product of the Arab sector was still larger in the 1920s, but by the 1940s it was only one-third of the product of the Jewish sector (see Table 4.3). The Jewish sector, which included barely over 30 per cent of Palestine's population in 1947, produced 75 per cent of the combined Jewish–Arab industrial product. Arab industry remained highly

Table 4.3 Palestine – Jewish sector and Arab sector industrial product, 1922–47 (1000s fixed 1936 PL)

Year	Jewish sector	Arab sector	Percentage of Jewish sector
1922	312	342	47.7
1927	693	749	48.1
1932	1,590	954	62.5
1937	3,356	1,524	68.8
1942	7,191	2,345	75.4
1947	9,984	3,323*	75.0

Source: Metzer and Kaplan 1990.

Note: * Computed on the basis of 1945 proportions.

concentrated in foodstuffs and textiles, whereas Jewish industry diversified into branches such as metalwork and jewellery–diamonds (see Table 4.4). Moreover, most Arab enterprises remained small-scale workshops and traditional crafts, whereas middle-sized plants were more and more common in the Jewish sector (see Table 4.4).

The Arab industrial sector probably benefited from the modernisation of infrastructure and administration by the British government, as well as from the rapid demographic growth of the Arab sector which had entered the second stage of the demographic transition model and enjoyed some in-migration. Nevertheless, the significant growth of industrial product in the Arab sector was associated with the trickle-down effects of capital and know-how

Table 4.4 Palestine – employed persons in industry and size of average plant by branch, 1942[a]

Industrial branch	Employed persons [b]			Average no. of employed persons per plant		
	Jewish	Arab	Total	Jewish	Arab	Total
Food	21.8	32.4	23.8	23	9	7
Chemicals	5.9	2.2	5.2	13	6	12
Rubber	0.7	0.0	0.5	14	–	14
Wood and cork	3.9	7.6	4.6	11	2	5
Paper, bookbinding, etc.	3.0	2.4	2.9	13	35	14
Textiles	9.0	20.1	11.1	20	8	13
Wearing apparel	14.0	15.1	14.2	15	4	10
Leather	2.6	2.4	2.5	17	5	12
Non-metallic minerals	5.4	3.1	4.9	33	5	9
Metals	21.5	12.9	19.8	21	5	15
Jewellery	9.0	0.0	7.3	95	0	95
Miscellaneous	3.3	1.9	3.0	16	15	16
Total (absolute no.)	37,773	8,804	46,577	20	6	13

Source: Gurevich *et al.* 1947, after Department of Statistics of the Government of Palestine, *Census of Industry*, 1943.

Notes: [a] The table does not include handicrafts and small-sized workshops, and cannot be compared directly with Table 4.1.
[b] Percentage of employed persons in industry out of total population group.

from the more rapidly growing Jewish sector (Abramovitch and Guelfat 1944; Horowitz 1946; Metzer and Kaplan 1990). Links between the Jewish and Arab sectors, however, were minimal: few Arabs at that time were employed in Jewish-owned enterprises, and the dual economic development has been characterised as 'joint but separate' (Metzer and Kaplan 1985). However, the 1936–9 Arab uprising terminated even the few surviving economic links. When the Arab boycott of Jewish products went into effect in 1946, thus blocking traditional Middle Eastern markets for Jewish industry, the separation of Jewish and Arab economies was completed.

SPATIAL TRANSFORMATION DURING THE BRITISH MANDATE ERA

Processes of spatial reorganisation of industry, consisting of a shift towards the coastal plain, began to occur during the early years of the British Mandate (Biger 1983). The central coastal plain, from the Haifa region in the north to the Tel Aviv region in the south, which formed the core of new Jewish industry, contained 50.8 per cent of industrial employment in Palestine as early as 1928. The three coastal towns of Tel Aviv, Jaffa, and Haifa alone comprised 25.1 per cent of the total number of plants, 40.5 per cent of total employment, 53.3 per cent of total industrial output, and 67.7 per cent of capital invested in manufacturing (see Table 4.5). Tel Aviv already had the third largest concentration of industrial activity, and, taken together with its neighbouring town of Jaffa, constituted at that time the largest industrial concentration in terms of employment, output, and investment (see Table 4.5). Tel Aviv formed an industrial base dominated by metalworks, textile, clothing, leather, and printing, whereas Haifa's industry was more oriented towards large food processing and building materials plants (Biger 1982). The Jerusalem area still had a substantial role in the industrial sector in 1928, particularly in stone quarries and other building materials, wood products, metalworks, jewellery, clothing, and printing. Traditional Arab industries dominated in smaller towns, with the continuation of particular specialisations in soap manufacturing in Nablus and weaving in Gaza.

Tel Aviv developed into the major industrial centre of Palestine mainly because individual entrepreneurs showed a strong preference for that city. This preference was due to several factors. First, Tel Aviv was established alongside Jaffa, the main and largest port

Table 4.5 Palestine – concentration of industry in the four largest urban centres, 1928 (percentage of total)

Location	Population	Industrial plants	Industrial employment	Industrial output	Industrial investment
Jerusalem	12.8	21.0	20.3	12.7	10.4
Tel Aviv	4.6	8.3	15.9	14.2	19.5
Jaffa	5.3	7.5	8.6	15.7	12.6
Haifa	9.2	9.3	16.0	23.4	35.6

Source: Biger 1982.

of entry to the country until the Haifa deep-water harbour was completed in the 1930s. This proximity to point of entry gave Jaffa and Tel Aviv an advantage for almost two decades, especially for the smaller-scale individual entrepreneur. Second, the town of Tel Aviv rose to prominence as the first Jewish town in the country, a fact considered as a locational advantage over the mixed Jewish–Arab town of Haifa by many of the newcomer Jewish entrepreneurs. Third, Jaffa and Tel Aviv were the gateways to the holy city of Jerusalem, which was also the largest urban centre until the 1930s and the centre of both British administration and Jewish organisations. While Haifa had a location of strategic importance in the Middle Eastern context, Jaffa and Tel Aviv enjoyed a favourable location from the point of view of Palestine's markets. As a rapidly industrialising town, and as a gateway to Jerusalem, Tel Aviv therefore gradually acquired the role of a major supplier of industrial goods to its industrially lagging, but administratively flourishing, counterpart city of Jerusalem.

Hence, Tel Aviv and its vicinity evolved as the major industrial centre, attracting light industries in particular. The dominance of Tel Aviv as a magnet for Jewish-owned industries grew with time (see Figure 4.2). During the boom period associated with the Second World War, the industrial concentration in Tel Aviv and its vicinity (including Petah Tiqwa) reached about 70 per cent of all Jewish-owned plants.

Haifa became a secondary industrial centre, specialising mainly in heavy industries. Its advantageous position in terms of sea and rail transportation was already evident during the last years of Ottoman rule. The British administration gave high priority to the development of Haifa Bay for strategic purposes, and decided to utilise it as their major port in the eastern Mediterranean area and as a gateway to the Middle East; consequently, Haifa was chosen as the site for the first deep-water port in Palestine and construction was completed in 1933. It also evolved as a railway junction, and railway repair shops and military bases were concentrated in and around the town. Several years after the completion of the port, Haifa became the outlet for the Iraqi crude oil pipeline and oil refineries were established (Sofer 1976).

These factors had an immense impact on the location of heavy industry. Proximity to the port of Haifa was essential for various modern heavy industries in order to save on the transport costs of imported raw materials. In fact, this perception of the port as a decisive location factor was probably more apparent than real (Sofer 1971). The oil refineries attracted more industries, and led in later years to the evolution of a petrochemical industrial complex. The immense industrial area planned in the 1930s in the southern Haifa Bay was influential in attracting space-consuming industries, with significant environmental effects.

The Jewish sector also seemed to earmark Haifa as a preferred location (Gross 1980). The town was selected as the site for the largest modern Jewish-owned industrial plants in the 1920s. This industrial structure, dominated by large enterprises, contributed to the strong tradition of organised labour in the area. Industrial plants established and acquired by the Federation of Labour mainly located in Haifa, where the power and official headquarters of the Federation were concentrated (Broido 1946). Haifa also became the centre for scientific and technological education. The Technion – Israel's technological institute – which was inaugurated in 1924, acted as an important locational magnet, explaining the relatively high concentration of electrical equipment plants in Haifa even in the pre-statehood years. However, despite these advantages and development, Haifa never managed to compete with Tel Aviv's supremacy, and the gap between the two cities tended to increase over time. This inferiority of Haifa was attributed largely to its inability to attract light industries (Sofer 1989).

Jerusalem, the site of the third industrial concentration, lost its supremacy to the vigorously growing coastal cities, and its industry remained dominated by small-scale crafts. Even the Second World War did not lead to substantial industrial growth in Jerusalem.

Whereas spatial statistics are not available for the Arab sector,

INDUSTRY

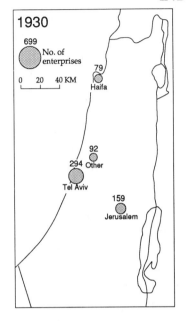

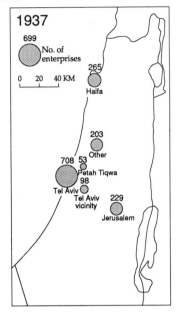

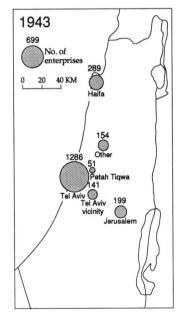

HANDICRAFTS

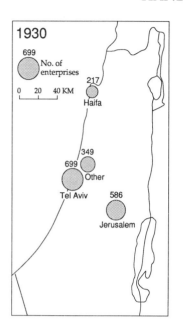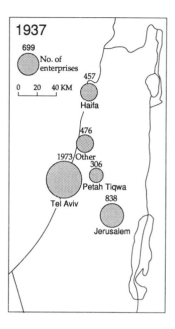

Figure 4.2 Palestine – spatial distribution of Jewish-owned industrial establishments, 1930–43

Source: Gurevich *et al.* (1947).

most evidence indicates that traditional local specialisations were retained to a large extent, although there was a certain shift towards the coastal plain (Abramovitch and Guelfat 1944). Nablus was still a major centre for olive-oil and soap production; Majdal, in the southern coastal plain, was a centre of fabric production; cigarettes were produced in Haifa and Jaffa; matches in Nablus; and, several relatively large metalwork plants were operating in Jaffa.

The 1948 War created a major transformation in Israel's/ Palestine's human geography. The State of Israel was established and gained control over the larger part of Palestine. Most of the Arab population within the boundaries of Israel abandoned their towns and villages, except for those Arabs in the Galilee, in the eastern edge of the central coastal plain, and some of the Bedouin in the Negev. The dual nature of economic development during the period of British rule explains why the practical abolition of the Arab industrial sector within the boundaries of Israel had little impact on the Jewish sector's industry, which thereafter became the Israeli industry.

5

EARLY CRYSTALLISATION OF THE ISRAELI SPATIAL INDUSTRIALISATION POLICY, 1948–55

Shortly after its establishment, Israel launched an extensive population dispersal policy motivated by defence and political objectives, as well as by such physical planning considerations as the reduction of congestion and the change of the primacy structure of the urban system (Sharon 1951; Brutzkus 1964). A major element of this policy was the establishment of more than thirty small- and medium-sized development towns, most of them during the 1949–56 period. The population dispersal policy, and in particular its urban element, achieved considerable success from a demographic point of view during the 1950s and early 1960s (Shachar 1971). This success can be attributed to the control the government was able to exercise over land, capital, and the immigrant population. Over 90 per cent of the land in Israel is controlled by the government. Private capital accumulation was still very limited in the 1950s and private businesses depended on the government for financial investments and for obtaining foreign currency. Government authorities financed between 50 to 80 per cent of gross capital formation in Israel between 1950 and 1958 (Patinkin 1960). Above all, the mass immigration of the 1950s and early 1960s could be directed with relative ease by the government to non-metropolitan development towns.

A systematic effort to disperse industry as a major means for providing employment in peripheral regions did not emerge until 1957. Previous calls for dispersion of industry in the 1940s and early 1950s were influenced by the 1940 report of the Barlow Commission published in Britain (Hall 1982), and stressed the disadvantages of large conurbations in terms of economic vulnerability and defence, as demonstrated in the depression of the 1930s and in the Second World War (Sharon 1951). These calls were also influenced by classical location theory, which classified manufacturing activity into raw materials-oriented, market-oriented and footloose categories, and emphasised the dispersal potential of footloose industries, which were considered to be insensitive to transport costs (Reichman 1979). However, these calls had little impact until the mid-1950s.

In contrast to valuable experience accumulated in the agricultural sector over a long period, no such body of experience existed at that time in the area of industrial and urban planning (Cohen 1970). There was no central authority for planning the new towns and coordinating implementation; nor did the Ministry of Commerce and Industry – very weak politically until 1955 – take part in the early phases of implementation of the development town policy (Beilin 1987). Thus, apart from a general perception that the new towns should provide central services to surrounding rural areas, the economic base of the towns did not receive careful consideration. In addition, scarce investment capital, directed towards agriculture and housing as a first priority, was not available for massive industrialisation projects (Brutzkus 1963; Eshkol 1963). What industrial development there was during the early 1950s was mostly in food processing, textiles, and construction-related industries, and remained concentrated along the central coastal plain. In 1952, the city of Tel Aviv accounted for 39.6 per cent of the industrial employment in Israel and the city of Haifa accounted for 16.8 per cent; the central coastal plain from the Haifa region in the north to the Tel Aviv region in the south, including these two cities, contained 87.4 per cent of all industrial employment in Israel (see Figure 5.1).

The major industrial projects in peripheral regions were carried out during the period 1950–7 by the government and by the Federation of Labour (Histadrut) enterprises. While the government began a wide-scale development of mineral extracting plants in the south (Razin 1985), Israel's General Federation of Labour corporations, officially committed to goals other than profit maximisation (Barkai 1983), played a central role in industrialising development towns. The large construction corporation of the Federation of Labour, Solel Boneh, and its industrial subsidiary, Koor Industries, were led by a powerful personality named Hillel Dan. Under his leadership, Solel Boneh evolved as the largest business organisation in Israel during the early 1950s. It was dominant in construction and in large-scale basic and heavy manufacturing, and became active in several development towns, mainly Akko in the North, and Ashqelon and Beer Sheva in the south (Biltski 1974).

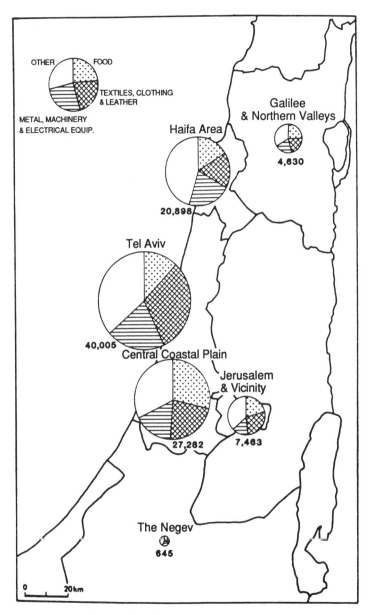

Figure 5.1 Israel – employed persons in industry by region and major branch, 1952
Source: Central Bureau of Statistics, *Census of Industry*, 1952.

The role of the private sector in industrial dispersal was minimal until the late 1950s. The new State of Israel, led by the Labour party, made a sharp transition from the pre-State system of *laissez-faire* into pervasive government intervention through austerity and price regulations. For a while, private industrialists feared that the Israeli government intended to follow the wide-scale nationalisation policy carried out by the Labour government in Britain after the Second World War. Emphasising the weaknesses of the British experience, private industry also tended to oppose calls for British-type industrial dispersal programmes that incorporated various controls on locating industry in major urban centres (*Hataassiya* 1949).

The Federation of Labour enterprises viewed themselves as the only ones capable of developing basic and large-scale mass-production industries enjoying economies of scale. They accused the private sector of being limited to small- and medium-sized traditional factories and of not contributing to the industrialisation of peripheral regions (Dan 1963). Officials of the Israeli Labour movement tended to make an analogy between Jewish agricultural development and the desired path of industrial development. Agricultural development was led by socialist cooperative settlements, which were supported by public funds, and the role of privately owned agriculture was limited and nearly confined to central regions. Acknowledging the limited scope of privately owned manufacturing plants, and following the rural example, the Federation of Labour corporations were assigned by Labour officials the role of leading development efforts in industry with the aid of public support (Beilin 1987).

Private industrialists complained bitterly that they were being discriminated against when compared with the Federation of Labour enterprises with regards to the supply of raw materials and public aid in financing capital investments. They argued that small, flexible and efficient plants were more appropriate for the size of Israel and for utilising the entrepreneurial skills brought by immigrants from the Diaspora. They accused the government of deterring entrepreneurship and discouraging foreign investors by the austerity programme and, later on, by leaving private industry in a position inferior to the huge Federation of Labour sector. Actually, due to the fact that private industrialists were risking their own capital, private industry was more cautious in locating outside the major urban centres (*Hataassiya* 1955). Political supporters of private industry therefore called for the development of industry without any spatial bias and recommended waiting until market mechanisms attracted private capital to the periphery (Beilin 1987). It should be stressed that this dispute in the early 1950s was conducted along ideological lines between socialist and capitalist-liberal approaches.

Since no explicit spatial policy of industrialisation existed during that period, early industrialisation of the development towns was accomplished through *ad hoc* decisions and was not based on formal criteria. Under these conditions, industrial development depended much on the initiative of local political leadership, and early calls for industrial dispersal tended to emphasise incentives offered by local government. A classical example of local initiative was the development town of Beer Sheva, where a homogeneous and relatively established group moved to the town and assumed responsibility for most municipal and regional positions and functions. This nucleus of leadership showed resourceful activity unconstrained by organisational frameworks and procedures. The first mayor, David Tuviau, attracted industry owned by the Federation of Labour to the town, as well as government investment in infrastructure and housing, mostly through his personal authority and contacts in the Federation of Labour as a former Solel Boneh senior employee (Ben Elia 1975). This initiative had a major impact on the rapid development of Beer Sheva in the 1950s.

The development town of Afula, in northern Israel, illustrates a different case. A small town for more than twenty years prior to the establishment of Israel, Afula was given an important role in relation to other completely new towns in early national planning, and a large number of new immigrants were settled there. However, the local leaders in Afula did not perceive economic development as one of their functions and limited themselves to the formal duties defined for local government. The passive attitude of the local leadership, which did not include dynamic personalities with strong positions in national political networks, caused Afula to miss the opportunity of receiving resources for development in the early 1950s. The central government interfered only in the late 1950s, after severe problems of unemployment and negative migration balance had plagued Afula for several years. Intervention by the government was thus too late and too little, since by the late 1950s many other development towns were competing with Afula for industrial investment. Afula lost its top-priority position and faced stagnation and continuous economic problems until the late 1970s (Ben Elia 1975).

6

THE GREAT PUSH FORWARD
Industrialisation of the development towns, 1956–67

REALISATION OF PRECONDITIONS FOR INDUSTRIAL DISPERSAL

Shortages in capital for investment in manufacturing were relieved during the mid-1950s, largely due to German reparation payments, which commenced in 1953 (Bank of Israel 1965), and to the shift in national priorities from agriculture to manufacturing (see Tables 2.1 and 3.1). Thus, an accelerated development of import-substituting industry began (Halevi and Klinov-Malul 1968). Complaints about discrimination against private sector industry intensified during these early years of rapid industrialisation. Hillel Dan, heading Solel Boneh, was appointed by the government to manage the company allocating the reparations. The reparations enabled vast industrial development of the Federation of Labour's Solel Boneh-Koor, some of which took place in development towns (Dan 1963; Biltski 1974).

Availability of capital was essential, but not sufficient, to activate wide-scale industrial dispersal. However, concurrently accumulated experience made it evident that the economy of the development towns could not depend to any large extent on providing central services to surrounding rural areas (Shachar 1971; Gradus and Stern 1980). Relief work and nearby agriculture served as temporary solutions, but a spatial industrialisation policy in the British post-Second World War planning tradition was increasingly perceived as the only long-term solution to the problems of the development towns. Hence, the Ministry of Commerce and Industry became involved with the population dispersal policy. A development regions unit was established and in 1956 a map of preference zones was drawn for the first time (Brutzkus 1963).

The commonly given explanations for the emergence of Israel's spatial industrialisation policy – availability of capital and accumulated planning experience – do not fully explain the precise form by which the new wide-scale incentive system, emphasising the private sector, evolved. The appointment of Pinhas Sapir as the Minister of Commerce and Industry in late 1955 was an important additional factor by itself. Sapir was the most clear-cut representative of institutional entrepreneurship in Israel; he managed Israeli industry, and later the whole economy, in a paternalistic manner. He excelled in the recruitment of funds from foreign investors and donors, and in encouraging private and public entrepreneurs and investors. At the same time, he did not contribute much to proper administration (Kimmerling 1983). This key figure stood behind the rapid industrialisation of the development towns during the period 1957–65.

The emergence of private sector industry in development regions also coincided with the curtailment of power of the Federation of Labour's major enterprise, Solel Boneh-Koor. The immense economic power accumulated by Solel Boneh during the 1950s alarmed senior officials of the Labour party, both in the government (including Sapir) and in the Federation of Labour itself. Hillel Dan of Solel Boneh acted more and more independently, disregarding officials of the Federation of Labour, who feared that the huge concern was beginning to get out of their control. Solel Boneh increasingly suffered from internal organisational problems and from industrial disputes with its workers, who enjoyed the support of the Federation of Labour. Arguments to the effect that Solel Boneh's enterprises enjoyed economies of scale, and that one should 'not clip the wings of the large eagle while in flight' were less convincing, since the conglomerate's rapid expansions had led to financial difficulties and pressures for government support. The Minister of Commerce and Industry and the Minister of Finance, however, did not relish the prospect of economic management by a public bureaucracy, and feared the negative influence of a widely nationalised economy on prospects for encouraging investment by Jews from the Diaspora (Schweitzer 1984).

In 1955, Hillel Dan was forced to resign from the company allocating German reparations, and in 1958, after a fierce struggle, the Federation of Labour decided to reorganise Solel Boneh, splitting it into three separate concerns. Accordingly, Hillel Dan resigned and the Federation of Labour enterprises, while remaining a most important element in Israel's economy, were never to regain their previous power (Biltski 1974). Koor Industries, no longer a subsidiary of Solel Boneh, was in a relatively weak position throughout the period of rapid industrialisation of the late 1950s and early 1960s, since it was unable to recycle profits from construction activities to finance industrial investment. Hence, private industrialists, who previously complained bitterly of being in an inferior position compared to the huge and politically powerful Federation of Labour enterprises, became prime beneficiaries of the government's incentive system for industrial dispersal.

THE GOLDEN YEARS OF INDUSTRIAL DISPERSAL – AN INTRODUCTION TO THE SAPIR ERA

The period 1956 to 1965 can be regarded as the 'golden era' of industrialisation in the development towns. The industrial basis characterising development towns for decades to come was laid down during those years. Industrialisation was led by the private sector, which received massive government support, and the percentage of industrial employment in the two peripheral regions – Northern and Southern – rose from 7 per cent in 1956 to 17.2 per cent in 1965 (see Table 6.1 and Figure 6.2). Preferred zones and development towns were defined, and criteria for receiving incentives – grants, subsidised loans, land, infrastructure, and tax exemptions – were gradually outlined (Zilberberg 1973). Nevertheless, Sapir still retained considerable flexibility for personal initiative, and he persuaded industrialists to invest in development towns and stepped in to rescue failing plants (Beilin 1987). Under these conditions, the possibilities for initiative on the part of local governments was substantially reduced (Ben Elia 1975), and this period can therefore be marked by an intensified strategy of development from above.

It should be noted that Israel's spatial industrialisation policy, formulated during those years, has never attained an independent status. There has been an industrial component in the population dispersal policy and a spatial component in the policy of encouraging capital investment. The industrial element did not have a central role in early phases of the population dispersal policy, since agencies in charge of this policy lacked authority in this field. The 1950 Law for Encouraging Capital Investment was intended originally to promote investment in the Israeli economy, and a clear spatial element was added to it only in the late 1950s, making it the major tool for dispersing industries up to the present day. Hence,

Table 6.1 Israel – employed persons in industry by district, for selected industrial branches and years[a] (percentages)

	District							Total in absolute numbers
	Tel Aviv	Central	Haifa	Jerusalem	Northern	Southern	Total (%)	
Industry (All)								
1956	49.6	17.0	20.3	6.1	3.9	3.1	100	93,791[b]
1958	48.6	17.6	19.3	6.3	5.4	3.0	100	117,863
1963	43.0	17.6	19.1	5.6	7.7	7.1	100	178,411
1965[c]	40.2	18.0	19.1	5.4	8.8	8.4	100	189,162
1967[c]	40.5	17.8	17.9	5.6	9.2	9.0	100	177,047
1972[c]	32.2	23.4	17.0	5.0	11.6	10.8	100	243,040
1977[c]	28.3	24.2	17.2	5.2	13.3	11.9	100	262,296
1983[c]	27.2	24.8	17.6	5.2	13.3	11.9	100	290,700
1985[c]	24.9	27.8	16.1	5.4	14.0	11.2	100	299,800
1988[c]	24.5	26.9	16.4	5.7	14.8	11.3	100	299,500
Diamonds								
1955	55.3	39.7	0.0	5.0	0.0	0.0	100	2,723
1958	52.1	34.9	0.0	5.8	5.1	2.0	100	3,476
1961	49.2	30.3	0.0	10.5	3.8	6.2	100	6,026
1963	57.7	27.1	0.5	7.0	4.1	3.7	100	8,046
1967[c]	58.7	29.7	0.8	4.0	2.4	4.4	100	9,272
1972[c]	58.9	31.2	0.0	6.1	1.0	2.7	100	8,696
1978	60.7	28.6	0.0	5.1	1.5	4.1	100	8,750

Table 6.1 continued

Textiles (old definition)							
1958	58.2	22.2	12.2	3.9	0.9	2.6	100	13,639
1963	45.2	14.4	8.8	3.8	10.8	17.0	100	22,522
1968[c]	41.8	12.2	10.3	3.2	13.2	19.4	100	28,300
(new definition)								+
1969[c]	42.4	10.9	9.8	2.5	11.6	22.8	100	29,347
1972[c]	21.3	19.7	13.0	2.9	15.2	27.9	100	24,167
1977[c]	18.2	23.6	13.6	1.8	17.1	25.6	100	20,983
1983[c]	22.4	22.8	15.6	1.2	10.8	27.5	100	16,700
1985[c]	29.2	20.1	6.9	2.1	13.9	27.8	100	14,400
1988[c]	25.4	21.5	6.9	–	16.9	24.6	100	13,000
Electric and electronic equipment								
1956	46.2	19.0	19.6	14.0	1.2	0.0	100	3,365
1963	65.1	12.2	16.7	4.5	1.0	0.4	100	8,010
1968[c]	55.3	15.9	22.5	3.2	0.9	2.1	100	11,143
1972[c]	43.0	25.4	16.8	2.3	2.3	10.2	100	20,440
1977[c]	26.8	44.4	15.2	2.8	2.5	8.3	100	26,246
1983[c]	22.2	38.5	27.7	1.5	3.0	7.2	100	40,500
1985[c]	22.6	36.5	27.1	3.2	3.8	6.8	100	47,100
1988[c]	21.6	35.4	24.4	4.7	6.6	7.3	100	42,600

Source: Central Bureau of Statistics, *Industry and Crafts Surveys*. Data for the diamond industry in 1972 and 1978 – Ministry of Industry and Trade.

Notes: [a] For district boundaries see Figure 6.1.
[b] Note directly comparable with 1985 data due to change in definitions.
[c] Data for plants with at least five salaried workers.

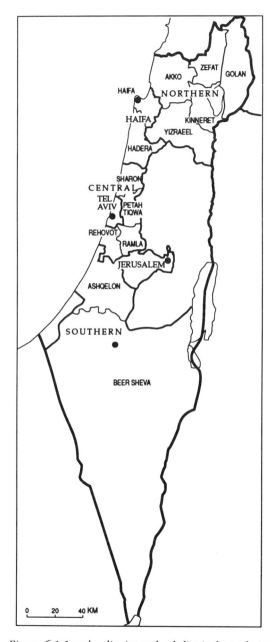

Figure 6.1 Israel – district and subdistrict boundaries

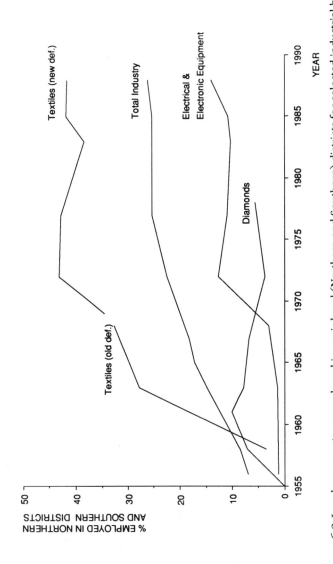

Figure 6.2 Israel – percentage employed in peripheral (Northern and Southern) districts for selected industrial branches, 1955–88

Source: Central Bureau of Statistics, *Industry and Crafts Surveys*.

industrial dispersal was not an independent policy, but an element of fluctuating importance within the more general policy intended to promote industrialisation. This seems to be a common feature of Israeli public policy, where attempts have often been made to catch two birds with one policy stone. Consequently, in times of budgetary constraints, the more pressing economic and political objectives would outweigh long-term spatial policies intended to reduce inequalities.

Until the mid-1950s, most industrialisation in the peripheral regions focused on raw materials or market-oriented activities (Zilberberg 1973). Nevertheless, planners emphasised the much greater dispersal potential of footloose industries, which were insensitive to transport costs of material inputs and outputs. Thus, since the mid-1950s the choice of industrial activities suitable for remote development towns has been based on two criteria: footloose industries, and industries offering ample employment for low-skilled workers, including women (Amiran and Shachar 1964).

EARLY FAILURE – THE CASE OF THE DIAMOND INDUSTRY

The first attempt to direct the private sector to remote development towns focused on the diamond industry. Israel emerged as a diamond centre during the Second World War, when occupation of Belgium and the Netherlands by German forces paralysed traditional diamond centres. Being a traditionally Jewish trade (Berman 1971), the young diamond industry in Palestine was established and reinforced by an influx of craftsmen and managers fleeing Europe. Plants were concentrated in the Tel Aviv area, in Netanya (located in the Central District), and, to a small extent, also in Jerusalem (see Table 6.2). After a series of reverses which led the Israeli diamond industry to almost total collapse—as the European diamond centres strove to recover lost ground—the diamond industry was entering a period of rapid growth in the mid-1950s (Szenberg 1971). The government had a central role in promoting expansion of the industry by assisting in raw material acquisition, credit, and manpower recruitment (Berman 1971). Experience gained during the Second World War, and the commercial and often familial and ethnic relations between local producers and marketing channels abroad, also facilitated the development of this industry.

The diamond industry was perceived to be especially suited for the development towns for several reasons: (a) capital investment

Table 6.2 The Israeli diamond industry – plants by major cities, 1946–72

	1946	1952	1961	1965	1967[a]	1972[a]
Tel Aviv	20[b]	66	127	204	139	143
Ramat Gan	–	4	19[c]	10	15	83
Bene Beraq	–	1	–	25	13	29
Petah Tiqwa	–	3	–	25	24	38
Netanya	11	21	67[d]	86	76	102
Jerusalem	2	7	27	20	13	17
Other localities	0	1	16	26	33	28
Total	33	103	256	396	313	440

Sources: *Hataassiya*, November 1946; Central Bureau of Statistics, *Industry and Crafts Surveys*; Central Bureau of Statistics, *Census of Industry and Crafts* 1965.

Notes: [a] Includes only plants with at least five salaried workers.
[b] Includes Tel Aviv's suburbs.
[c] Includes Bene Beraq.
[d] Includes Petah Tiqwa.

necessary for establishing diamond-polishing plants was relatively modest; (b) jobs offered by these plants were most suitable for the local labour force in that they did not require previous skills while offering satisfactory skilled-worker income after a relatively short period of experience; and, (c) transport costs were negligible (Danieli 1958; Szenberg 1971). The government offered subsidised credit and vocational training, with the result that fifteen diamond-polishing branch plants were set up during the period 1956–8 in development towns from Qiryat Shemona in the north to Elat in the south (Ministry of Commerce and Industry 1960). In 1961, 10 per cent of the employees in the diamond industry were in the Northern and Southern Districts (see Table 6.1), polishing mainly relatively cheap and standardised stones.

The performance of the diamond industry in development towns soon proved very disappointing (Ben Moshe 1977), and most early plants closed down after only a few years of operation (Razin 1985). Early reports traced the failure of these dispersal efforts to the lack of financial reserves and administrative expertise of owners, and to difficulties experienced in absorbing the new immigrant labour force (State Comptroller of Israel 1963). Examination of early failures led, therefore, to a revision of criteria for assessing applications of investors, and to modifications in vocational training assistance. An additional response to the crisis of the diamond industry was the decision to concentrate it in select development towns, where it was hoped that the polishing plants would be able to centralise buying, selling, and services (Ministry of Commerce and Industry 1960).

In spite of these moves, the polishing industry in peripheral regions entered a period of stagnation and decline in the early 1960s (see Table 6.1). Being an export industry, it was prone to frequent shifts in the world market for polished diamonds, and its growth prospects narrowed in the 1960s when Israel reached about 30 per cent of both the total value of world exports and employment in this industry. The Israeli diamond industry has not subsequently surpassed employment peaks reached in the mid-1960s. Branch plants in development towns tended to be the first ones to be closed down during the frequent downturns in this industry. Of all industries, it suffered from the highest closure propensity in the development towns (Razin and Shachar 1987). Remnants of the diamond industry continued to be operated by local residents of some development towns, mainly Zefat in the north, and Netivot and Ofaqim in the south, who acquired skills in the branch plants of the late 1950s and early 1960s, and these have survived as small and unsophisticated subcontractors (Ben Moshe 1977).

Meanwhile, the diamond industry was becoming even further concentrated in the Tel Aviv district, particularly in Ramat Gan (see Tables 6.1 and 6.2). This was due to the opening of the new diamond exchange in Ramat Gan in 1968 – where all trading operations were centralised – and also to restructuring processes, which led to the rationalisation of some large plants, and the vertical disintegration of activities into smaller, more flexible units (Ministry of Industry and Trade 1986). The severe recession in the diamond industry during the early 1980s dealt another blow to this industry in the development towns, leaving only a few surviving plants.

Despite all evidence, assistance for establishing new diamond-polishing plants in peripheral areas continued to be disbursed throughout the 1960s and 1970s. Even during the 1980s, it has still been claimed that Sapir should have given more emphasis to the dispersal of skill-intensive diamond export plants over textiles and plywood factories (Schweitzer 1984). However, in retrospect, it seems that the major assumptions underlying the dispersal efforts of the diamond industry were in error, and it has been suggested that the inability to benefit from external economies in this sector was the root of the problem (Szenberg 1971). Clearly the conception of transport costs of materials as a major location factor, while in line with classical location theory of the early twentieth century, was a major error. In the case of the diamond industry, additional transaction costs associated with each output linkage were of much greater importance, owing to the complex patterns of small and irregular linkages in this industry. These linkage patterns are associated with a complex subcontracting network (Szenberg 1971), and with a unique mechanism of trade which has traditionally been based on trust, oral agreements sealed with a handshake, and tightly-knit human relationships (Berman 1971). Plants having such linkage patterns tend to locate in densely developed complexes near major markets or centres of trade (Scott 1988). The planners were unaware that diamonds present a classical case of an industry tending to spatial concentration. The failure of this sector in the development towns was, therefore, inevitable due to the physical distance from management and trade centres in the Tel Aviv metropolis, particularly the diamond exchange centre in Ramat Gan.

SPATIAL DISPERSAL LED BY THE TEXTILES SECTOR

Major efforts were focused after 1957 on dispersing textile plants as a leading industry in the periphery. The textile industry was a traditional Jewish sector for generations and one of the better-established industries in the Israel of the 1950s. Textile plants in development towns were expected to provide a market for cotton from nearby agricultural areas. In principle, however, they were generally footloose, with location unaffected by transportation costs of raw materials and final products. Most important, the industry was labour-intensive, generally offering unskilled employment needed in the development towns (Toren 1979).

Textile plants heavily financed by the government evolved as a dominant industry in many of the development towns. These towns attracted mainly large, externally controlled, capital-intensive, vertically integrated plants, which were less dependent on complex networks of input and output linkages, and which benefited most from the government's capital subsidies. The Northern and Southern Districts' employment share in the textiles sector increased from 3.5 per cent in 1958 to 27.8 per cent in 1963, and to over 40 per cent in 1972 (see Figure 6.2; also Table 6.1).

During the late 1950s, textile industrialists also had hopes that some development towns would evolve as 'company towns', where employees would be linked for life with a large dominant enterprise which would benefit from a low rate of employee turnover and lower labour costs (Klir 1957). These expectations were not realised (Shinan-Shamir 1984). Industry in development towns suffered from social problems inherent in the new immigrant population, while nation-wide wage agreements kept wages at levels differing little from those in the central regions. The new plants suffered from high turnover of employees, and those obtaining adequate qualifications and experience tended to leave after a while to obtain better jobs in the central area of the country, thus creating a continuous need for training new personnel (Ministry of Commerce and Industry 1964). The large textile plants suffered frequent crises, resulting primarily from mismanagement and labour relations. Nevertheless, most plants managed to survive with the support of the government. Employment in the textile sector was generally attractive when compared with the relief work offered by the government or seasonal agricultural work. Thus, the textile plants continued to serve as major job providers and played a major role in solving employment problems in development towns until the 1970s (Zilberberg 1973; Razin 1985).

Other industrial activities were also emerging in development areas aside from textile plants. The most notable large-scale employers were the agricultural processing industrial complexes established by the rural-cooperative sector, which had a major role in solving employment problems in towns such as Tiberias, Bet Shean, and Qiryat Shemona in the north and Sederot in the south (Brutzkus 1963). These plants were usually characterised by high levels of capital intensity due to the extent of government subsidies. The rapid industrialisation of the periphery led to an increasing concentration of large plants in these regions: in 1958, although only 5.8 per cent of the total number of manufacturing plants were located in the Northern and Southern Districts, these districts included 12.5 per cent of all plants employing over 300 workers. The absolute share of the Northern and Southern Districts increased only to 7.8 per cent by 1963, but 26.8 per cent of the total number of plants employing over 300 workers were already located there (Central Bureau of Statistics, *Industry and Crafts Surveys*).

The government also continued to invest directly in industry in the development regions, mostly through vast expenditures to expand the mineral-extracting plants in the south (Razin 1985).

Additional efforts were made by the government through partnerships with the Federation of Labour and the private sector. A concern named Teus Development Areas was established in 1958 by the government and the Federation of Labour to initiate dozens of small- and medium-sized plants in various industries. The intention was to sell these plants or to transfer them to cooperatives of workers after a period of training and consolidation. This project met with little success, and, after years of successive losses, the government transferred its shares in 1967 to the Federation of Labour, which continued to operate the concern on a more profitable basis. Development corporations were established in order to assist the economic-industrial development of specific development towns. These were usually joint ventures of the government, the Federation of Labour, and the local authority (Beilin 1987). Involvement of the government in rescuing failing plants led in several cases to the transfer of shares to the government. However, direct governmental involvement in owning and managing industrial plants in the periphery led to heavy losses. The government retreated from its direct action in the periphery during the late 1960s, except for its mineral-extracting plants in the south and the defence industry plants, which began emerging in peripheral areas.

The period of rapid industrialisation in development towns came to a halt during the recession of 1965–7. Development towns were hard-hit by the recession, due to their dependence on a very narrow economic base consisting of manufacturing and construction, both of which were most sensitive to economic fluctuations. The small share of self-employed in the economy of development towns, and the large numbers of residents working out of town, also contributed to their instability (Don and Bar-El 1972). The recession, while a traumatic event for residents of development towns, had only a short-term economic impact. A rapid recovery process began shortly after the 1967 Six Day War which had a much more lasting impact on industrialisation trends in development towns.

IMPLICATIONS FOR INDUSTRIALISATION OF THE LARGE CITIES

The Israeli spatial industrialisation policy comprised mainly the positive ('carrot') elements of British spatial policy (Hall 1982; Law 1985). Negative ('stick') measures of restraining metropolitan growth were more difficult to apply (Shachar 1971). Practically, since 1965 the main growth-constraining element has been a statutory committee veto power over urban projects planned on agricultural land. This measure restricted to some extent the expansion of space-intensive industries in the Tel Aviv and Haifa metropolitan areas.

Haifa suffered more than Tel Aviv from competition exerted by the development towns, being, on the one hand, closer to development regions entitled to incentives, and, on the other, remote from the major Tel Aviv market and major business and government control centres (Biltski 1981; Sofer 1971). The Ashdod deep-water port, opened in 1965, eroded Haifa's hitherto exclusive advantage in this area. The relative stagnation of Haifa's economy resulted also from its reliance on basic and heavy industries. These industries had not experienced rapid growth since the 1960s, and some of them moved out of Haifa to the development towns of Akko and Beer Sheva.

Political circumstances provoked a sharp decline in investment in Federation of Labour industries in Haifa. Haifa was the powerful base of the Solel Boneh organisation, and the reorganisation of this conglomerate in 1958 was fiercely resisted. The political power struggle that emerged from this restructuring process had far-reaching spatial repercussions. The headquarters of Koor and of most other enterprises of the Federation of Labour subsequently moved to Tel Aviv, and the share of the Haifa subdistrict in Koor's employment declined from 53.2 per cent in 1954 to 21.5 per cent in 1969 (see Table 6.3). Moreover, Haifa was also the seat of a powerful local workers' council responsible for several major labour disputes during the 1950s and 1960s, thus earning the image of a problematic location in terms of labour relations. This unattractive business climate deterred both private investors and Koor, whose relations with the Haifa Workers Council were strained (Biltski 1981).

The decline of Haifa's industrial base was slow (see Table 6.1), and the city retained stability and high wages throughout the 1960s and later (Shefer 1972). Nevertheless, while the industrial base of Tel Aviv appeared to decline much faster (see Table 6.1), some of these statistics merely represent the suburbanisation of economic activity to the Central District, and the Tel Aviv economy flourished as a finance and control centre (Shachar 1974), and as a centre of small-scale manufacturing and service businesses.

Significant public attention was given during the 1950s to the

Table 6.3 Employed persons in plants owned by Koor Industries, by subdistrict, 1954–80[a]

Subdistrict	1954	1958	1969	1980
Zefat	0.0	0.2	1.3	1.7
Kinneret	0.0	0.0	0.0	0.0
Yizreel	0.0	3.7	9.2	5.5
Akko	8.5	13.1	11.0	8.8
Haifa	53.2	41.0	21.5	15.5
Hadera	0.0	0.0	0.0	7.1
Sharon	0.0	0.0	1.0	0.9
Petah Tiqwa	0.0	4.3	2.4	11.4
Tel Aviv	10.4	12.7	26.7	22.0
Ramla	7.2	6.0	11.6	6.7
Rehovot	1.2	0.0	0.0	4.0
Jerusalem	0.5	2.4	1.5	2.1
Ashqelon	15.6	11.8	6.7	4.9
Beer Sheva	3.3	4.8	7.0	9.1
Devel. zones – total[b]	27.9	36.0	36.7	32.2
Devel. zones – total[c]	12.4	24.2	30.0	27.7
Total	100%	100%	100%	100%
Total (absolute no.)	4,607	6,139	10,405	27,910

Sources: Razin 1984; 1991b.

Notes:
[a] The table includes plants in which Koor held at least 50 per cent of the shares, and does not include Koor's head office and non-industrial subsidaries
[b] Development zone according to the 1967 definition (see Figure 11.1a)
[c] Development zone according to the 1972 definition (see Figure 11.1b)

need to strengthen the economic base of Jerusalem. In addition to being land-locked, Jerusalem was surrounded by a closed border on three sides as a result of the 1949 ceasefire, and also lacked international recognition of its position as the capital of Israel. However, despite the establishment of a public economic corporation for the development of Jerusalem, and the provision of infrastructure for industrial development, local industry was slow to develop. Jerusalem suffered until 1967 from scarce land, an inferior location, and no tradition of modern industrial development. As efforts to industrialise the development towns intensified during the late

1950s, attempts to industrialise Jerusalem, which did not suffer from significant unemployment, dwindled. Even the few head offices of government-owned industrial corporations gradually left the city, and the development of Jerusalem focused on public administration – mainly government offices – and the Hebrew University.

7

THE POST-1967 CROSSROADS

The Six Day War had far-reaching implications for the Israeli economy, in particular stimulating a growth of military and high-technology industries which affected Israel's industrial geography. These long-term implications have been fully realised only since the 1970s and are discussed in Part III. Hence, this chapter focuses only on the direct consequences for the industrial dispersal policy, which was formulated during the decade preceding the war.

The Six Day War and the years of economic expansion that immediately followed were a turning point for the development towns, although the crucial importance of this period was to be recognised only years later. The principles of the spatial policy of industrialisation stayed in effect and even received further formalisation when an official map of preferred zones became statutory for the first time in 1967. This map was modified in 1972 (see Figure 11.1a–b; Zilberberg 1973). Programmes for population dispersal also attained statutory status in 1975 (Reichman and Sonis 1979). The recovery from the 1965–7 recession led to further dispersal of industry to peripheral regions, although at a slower rate (see Figure 6.2 and Table 6.1). Nevertheless, the institutional entrepreneurship of previous years was gradually replaced by more formalised government bureaucratic procedures. Such a transition may have been a natural evolution, assuring continuation of the policy in the long-run. However, while external and internal conditions were changing, it became increasingly difficult to substantially modify the policy. New national goals emerged, and while advancement of the development towns remained a goal in itself, its significance was somewhat eroded by the presence of new competing goals.

The first competing goal emerged as economic policy became increasingly export-oriented. Centrally located exporting plants were entitled to most government incentives under the Law for Encouraging Capital Investment, and thus benefited little from moving to development towns in order to attract public subsidies (Schwartz 1985). Development towns therefore attracted mainly non-exporting plants, whose sole eligibility for government incentives was due to their location, and particularly those traditional 'laggard' industries that found it most difficult to be eligible for incentives in central regions. Multiplant firms tended to take particular advantage of this situation by locating units in development towns that produced for the Israeli market (Razin 1988a; 1988b).

The development of an import-substituting capacity in military products evolved as another major Israeli national goal after 1967 (Zilberberg 1973; see also Table 2.1). This was mainly a result of the French arms embargo of the Middle East, which was imposed in 1967. Hence, firms related to military production were entitled to government assistance even if located in central regions. A striking case was that of the two large government-owned enterprises – Israel Aircraft Industries and Israel Military Industries – which grew immensely after 1967, until, in 1979, they employed about 34,000 workers. This growth, however, contributed relatively little to the economy of the development towns. These firms did not face problems of financing investments in central regions, so that location in a development town offered no significant advantage (Razin 1988b; Razin and Shachar 1990). Thus, the emerging goals of export promotion and import-substituting in military manufacturing meant that industries in the most advanced and expanding sectors were also entitled to most capital incentives in central regions.

The early phases of settlement in sparsely populated areas of the occupied territories constituted a third emerging goal. Planning efforts and institutional initiative increasingly focused on the establishment of settlements in these areas (Harris 1978). Major efforts were also undertaken to encourage Jewish settlement in eastern Jerusalem in order to assure its unified status within the State of Israel (Benvenisti 1976). While none of these efforts yet competed for resources for industrial development, they did divert the energy of 'doers' in the Labour party – which was still in power – from the traditional spatial goals of developing the Galilee and the Negev to new and more exciting spatial ventures.

Redirection of effort to these new national goals had an inevitable influence on the development towns. Industrial dispersal continued, but at a slower pace, and renewal of the industrial character of the development towns did not take place. Statistical analysis of the distribution of industry for the years 1965, 1971, and 1978 indicates that industrial activity in development towns was characterised by a persisting low degree of diversification, creating a dangerous dependence on a limited number of industrial branches (Gradus and Eini 1981; Gradus and Krakover 1977). The periphery was dominated by large capital intensive plants in 'laggard' industries (Kipnis 1977; Efrat 1977). The spatial industrialisation policy was successful in solving acute unemployment problems in the development towns but, resembling the experience of several European countries, fell short of creating self-sustained growth in the problem areas (Law 1985). The persistent interregional gap (Lipshitz 1986) led to high negative migration balances (Shachar and Lipshitz 1980) and to a gradual re-emergence of unemployment problems.

The stagnation of the industrial base of development towns faced both increasing external restructuring processes and gradual internal social change. External restructuring processes meant that many of the 'old generation' industries, such as textiles and those related to construction, which had dominated Israeli industry during the 1950s and 1960s, were beginning to contract as a consequence of heavy international competition and saturation of internal markets. However, these old generation industries continued to predominate in the development towns. By contrast, the major emerging industries of the 1970s – high-technology electronics and military-related industries – were not encouraged to enter the development towns since they were not affected by the spatial incentives incorporated in the Law for Encouraging Capital Investment (Felsenstein 1986). Some movement of intermediate growth industries such as metal products, chemicals, and plastics, to the development regions took place during the 1970s. Nevertheless, stagnation became more and more evident, particularly in the Southern District, where the crisis in the textile industry was not counterbalanced by the rapid growth of other industrial branches (Gradus and Eini 1981).

During the same period, development towns were characterised by gradual social change as second generation immigrants began entering the labour market. This new generation raised in Israel had attained higher skills and had higher expectations than their parents, and thus preferred, in many cases, to either migrate or receive unemployment benefits over work in low-wage monotonous jobs in old generation plants. The local industry, facing labour shortages, reacted by employing increasing numbers of Arabs from remote locations (Shinan-Shamir 1984; Bar-El and Schwartz 1985).

In the advent of the 1967 war, the Arab workers, particularly residents of the occupied territories, have gradually formed a pool of cheap unorganised labour (Lewin-Epstein and Semyonov 1986), resembling in many ways the foreign guest workers in Western Europe (Salt 1981; Castells 1975), or the Third World immigrant labour in large American cities (Portes and Bach 1985). These Arab labourers from the occupied territories penetrated and gradually dominated jobs at the bottom of the wage/prestige ladder, allowing for the upward mobility of previously low-paid Jewish workers, who were mainly immigrants from Middle Eastern countries. The Arabs particularly dominated unskilled seasonal jobs in agriculture, construction and service industries, but also penetrated low-level employment in most sectors of the economy, including manufacturing (Portugali 1989).

As residents of the occupied territories, and due to Arab/Jewish tensions, Arab workers were forbidden to stay overnight in Israel. While this measure was not totally effective, it created huge daily and weekly commuting flows from the Gaza Strip and the West Bank to the major Israeli urban centres. Around 45 to 50 per cent of the Arab labour force in the Gaza Strip and 30 per cent of the Arab labour force in the West Bank worked in Israel during the 1980s. Arab workers from the occupied territories have not been organised or protected by trade unions. Only 45 per cent of them have been employed legally through government employment agencies, leaving more than one-fifth of the Arab workers to find work through street labour markets at road junctions or on street corners, which serve as major informal channels for the supply and allocation of Arab labour. Others have obtained jobs through private Arab labour contractors or private labour agencies (Portugali 1989).

The main beneficiaries of employing such labour have been small firms, as it has provided them with greater flexibility in hiring, firing, and changing working conditions (Razin 1988b). Nevertheless, large mass-producing plants of big firms have also been compelled to employ an increasing number of non-local Arabs due to the reticence of local labour in the metropolitan areas, as well as in development towns, to work in low-paying traditional manufacturing. In the Galilee, where a large number of Israeli Arabs reside, the Arab minority also has tended to fill low-wage industrial positions, although not nearly under the same poor conditions as Arabs from the occupied territories. Minorities and residents of the occupied territories constituted nearly 22 per cent of the industrial labour force in eight medium-sized development towns surveyed in 1984 (Razin 1986). In the Maalot, Tefen, and Carmiel industrial zones, established to provide an economic base for two development towns and nearby Jewish settlements, 34.3 per cent of the workers were Arabs from nearby and distant localities (Yiftachel 1991).

As in the case of the guest workers in Western Europe, certain types of occupations have come to be reserved for, as well as associated with, the Arab labour force, since the indigenous Jewish labour force has sought to find alternative employment. The availability of relatively cheap Arab labour has discouraged the upgrading of capital stock and has further increased the bad image these jobs have gained (Shinan-Shamir 1984). Spatial industrialisation policy has not reacted to these changing labour market conditions, and thus has been unable to prevent the re-emergence of unemployment and outmigration problems in the development towns.

Part III

THE INDUSTRIAL GEOGRAPHY OF ISRAEL IN A PERIOD OF ECONOMIC STAGNATION

8
CHANGING REALITIES OF THE 1970s AND 1980s

The evolution of the industrial geography of Israel since 1973, its present patterns and prospects for future change, should be viewed in the context of the deepening economic stagnation which has reduced the pace of spatial-industrial change in Israel. Post-1973 external economic shifts, together with earlier shifts originating in the post-Six Day War period, have affected the performance of the national spatial industrialisation policy. These changing realities of the 1970s and 1980s can be classified by four major factors:

1 *Shifts in economic policy.* Growing role of exporting industries and military import-substituting industries, which have constituted new national economic goals since 1967.
2 *Political and labour market implications of the Six Day War.* Occupation of new territories, leading to the emergence of new national geo-political goals, and to the incorporation of Arab labour from the occupied territories into the Israeli labour market.
3 *Changing external economic realities.* Reduced growth rates in the Israeli economy since 1973, and shifts in global economic conditions associated with the restructuring of industrial activities.
4 *Changing social-political conditions in the development towns.* The reduced compatibility of local industrial activity to the skills and aspirations of the second generation development town inhabitants entering the local labour force.

Incentives offered for the industrialisation of development towns remained in effect throughout the 1970s and 1980s, with minor changes in terms of loans, grants and export requirements (Schwartz 1985). By the 1980s, it became clear that the spatial policy was at an impasse. Formulated during the 1950s, it has been unable to adjust to new realities that have demanded the revision of the map of government assistance; a reorientation of public policy towards new types of industries and enterprises; the decentralisation of government agencies engaged with regional development; and an update of the priorities for allocating public resources for development of peripheral areas. This inability to adjust can be traced to four factors: a significant time lag in identifying external changes and in adapting new ideas and planning perceptions; a general weakness of central government when confronting strong interest groups that benefit from the preservation of the present policy; bureaucratic inertia and the gradual formulation of statutory bureaucratic procedures that eventually serve to hamper attempts for change; and, stagnation of Israel's economy which blocks efforts to strengthen what can be considered part of the welfare state mechanisms established during earlier periods of rapid growth. Industry has suffered from sluggish growth (Bar 1990), and the lack of significant growth has meant that development areas have been competing in an increasingly zero-sum game against central regions over industrial investment. Hence, spatial industrialisation policy has survived mainly as a tool for alleviating short-term unemployment problems and has not pulled the development towns into a course of long-term growth. Since 1977, the share of the (Northern and Southern) peripheral districts in Israel's industrial employment has not grown (see Figure 6.2; see Table 6.1).

The re-emerging economic problems of development towns have led to intensified criticism of the spatial policy. One critique has focused on the early roots of the policy, as formulated during the first years of the state. Implementation of this policy in the early days led, according to this argument, to persistent problems of interethnic and spatial inequalities. Referring to dependency theory, such an argument stresses the discriminatory means by which new immigrants, mainly from Middle Eastern and North African countries, were absorbed in Israeli society by the existing population of mostly European origin. These practices have led to the long-term preservation of disadvantage of the Middle Eastern and North African immigrants in Israeli society, and particularly to the lasting disadvantage of development towns inhabited mainly by these immigrants (Svirski and Shoshan 1985). Other critical appraisals of the population dispersal policy focused on physical planning misconceptions, such as those which led to the establishment of too many small development towns rather than fewer larger towns (Efrat 1987).

As to the spatial industrialisation policy, the actions taken by Pinhas Sapir during the late 1950s and early 1960s have been extensively criticised. It has been argued that Sapir directed laggard industries to the development towns, and that he actively discouraged local entrepreneurship, built private capitalists through the transfer of public funds, and generally increased the dependency of entrepreneurs on the government (Schweitzer 1984). In light of the details presented in this book, it is our view that such assertions are not sufficiently proven. At the micro level, it is easy to identify mistakes originating from the non-selective encouragement of investments and from other factors. The negative implications of the paternalistic mode of action practised by Sapir are also quite clear. However, with hindsight, it is difficult to point to an alternative strategy that could have been implemented under the conditions of that period and which would have led to better results in the development areas. Arguments of Schweitzer (1984) that Sapir should have emphasised the dispersal of skill-intensive export diamond plants over textiles and plywood are clearly erroneous, as is shown in Chapter 6. At this juncture, we argue that the major mistakes and oversights of the spatial industrialisation policy in fact occurred after the Six Day War, when policy makers retained past practices and disregarded changing realities.

The following seven chapters focus on the major processes influencing Israel's industrial geography during the last two decades of economic stagnation, and on their spatial implications. Chapter 9 presents the results of a long-term quantitative assessment of the industrial dispersal policy in order to set the stage for the discussion which follows. Chapter 10 deals with the ongoing economic problems of the periphery, and with the mostly futile calls for revising priorities and diversifying the industrial base of the periphery. These calls were largely made too late, and the remedial recommendations were mostly based on past conditions that were no longer relevant. Chapter 11 examines the direct impact of the government's incentives on the spatial behaviour of industrial enterprises. It evaluates the incremental modifications in the policy and stresses the much-overdue need for revision. The impact of mass immigration at the onset of the 1990s on dispersal policy is also assessed. Chapter 12 relates to corporate geography. It emphasises the spatial implications of the major crises and decline of the Federation of Labour enterprises, reflecting the way in which structural-organisational change has influenced Israel's industrial geography.

Chapter 13 examines the impact of the major agent of structural change in Israel's industry during the last two decades – the growth of high-technology industries. It emphasises their strong clustering tendencies, and the attempts at their dispersal.

Chapter 14 deals with the shift in balance between central government and local authorities in the realm of economic development. It examines factors encouraging the emergence of locally initiated development policies, describes competing strategies, and evaluates them and their spatial consequences. Chapter 15 treats the shifting trend in planning and policy-making circles towards promoting local entrepreneurship. The problems associated with initiating such a strategy are described and possible regional implications are assessed.

9

MEASURING THE EFFECTIVENESS OF INDUSTRIAL DISPERSAL POLICY
A quantitative assessment

The effectiveness of industrial dispersal policy in Israel has been analysed in several studies, most of which have utilised the official division of the State of Israel into the fourteen subdistricts shown in Figure 6.1. These subdistricts have been generally classified into three groups: the two industrial core units of Tel Aviv and Haifa; a group of six centrally located subdistricts ranging from Hadera in the north to Rehovot in the south, including Jerusalem; and a group comprising six peripheral subdistricts – four in the north of the country and two in the south.

LEVEL OF INDUSTRIALISATION

Using this geographical division, Gradus and Krakover (1977) examined changes in three industrial dispersal indices over the period 1965 and 1971. Figure 9.1a exhibits the level of industrialisation reached in 1965 through the use of a location quotient (LQ) that measures the ratio of employment in manufacturing industry to total employment in each subdistrict. By comparing the LQs for 1971 with those for 1965, Gradus and Krakover were able to show that the level of industrialisation was on the rise in all peripheral subdistricts (see Figure 9.1b). Furthermore, during the same period, industrialisation was shown to have declined in the two industrial core subdistricts of Tel Aviv and Haifa. Results in the central area were mixed; levels of industrialisation increased in three subdistricts and declined in the other three. Of the central subdistricts, only the growth in Ramla was attributed to the process of suburbanisation of industrial activity. This growth was probably due to the expansion of the Israel Aircraft Industries complex located in the vicinity of the Ben-Gurion International Airport.

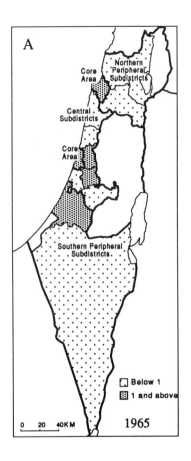 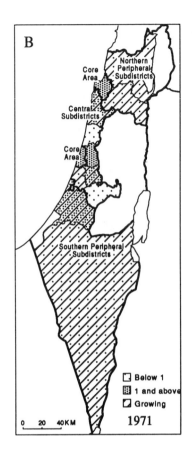

The overall results obtained for the period between 1965 and 1971 indicated that the policy of industrial dispersal introduced more than a decade earlier has helped to initiate a trend in the expected direction. The application of this policy contributed to the generation of industrial jobs in the periphery at a greater rate than in the central parts of the country. A follow-up study by Gradus and Eini (1981) found, however, that up to 1977 some of the peripheral subdistricts were losing ground, especially in the south (see Figure 9.1c). The central parts of the country once again exhibited mixed results: Tel Aviv and two of its neighbouring subdistricts were

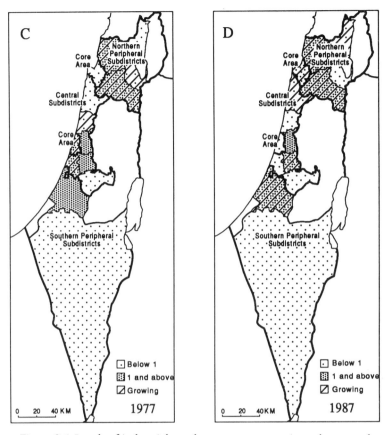

Figure 9.1 Levels of industrial employment concentrations, (measured by LQ), 1965–87

gaining relative concentrations of industrial employment, while Haifa and the rest were declining.

A recent examination of the levels of industrialisation for the year 1987 has revealed that the three peripheral subdistricts adjacent to the central region have exhibited relative industrial employment concentration (see Figure 9.1d). On the other hand, of the three more remote peripheral regions, only one has experienced relative gain. This geographic distribution is attributable to the behaviour of private investors who tend to exploit the existence of government financial incentives for those areas adjacent to the central districts of the country. This pattern was avoided two decades earlier when government discrimination in favour of the peripheral areas was heavier and more direct.

A long-term examination of the levels of industrialisation portrays several longitudinal patterns (see Figure 9.2). Two of the northern peripheral subdistricts increased their level of industrialisation constantly between 1965 and 1987, and have been above the mark of 1.0 LQ during the last ten years. Both of them are conveniently located in adjacency to the core region of Haifa. A third peripheral region, conveniently located south of Tel Aviv, always showed the highest level of LQ among the peripheral subdistricts. The three more remote peripheral subdistricts were fluctuating in their level of industrialisation, but never exceeded the 1.0 LQ mark (see Figure 9.2c). This classification seems to justify subdividing these six subdistricts into three semi-peripheral and three entirely peripheral subdistricts (see Figure 14.1).

The two core industrial areas of Tel Aviv and Haifa exhibit long-term industrial employment decline, though a resurgence of industrial growth is evident in the latter in the last ten years (see Figure 9.2a). Among the six central subdistricts, two are constantly above the 1.0 LQ mark, while the rest have approached unity LQ in the past but were not able to retain this level over time (see Figure 9.2b). One exception is the Jerusalem subdistrict, which declined to the lowest level of industrialisation among the central subdistricts.

The overall longitudinal pattern of spatial dispersion of industries that emerges from this analysis is one of moderate decline in the core areas and industrial employment growth that, in general, bypasses the central subdistricts in favour of the semi-peripheral subdistricts. The reasons for this pattern are elaborated in Chapters 11, 13 and 14.

LEVEL OF DIVERSIFICATION

A second measure of industrial development used by Gradus and Krakover (1977) indicated that industrialisation of the peripheral areas was achieved in the 1960s through declining diversification. Levels of industrial diversification were measured by applying the Lorenz curve and calculating its Gini coefficient. A follow-up study by Gradus and Eini (1981) showed, however, that by 1977 levels of industrial diversification had increased in five of the six peripheral subdistricts. This was interpreted as a further success of the national

THE EFFECTIVENESS OF INDUSTRIAL DISPERSAL POLICY

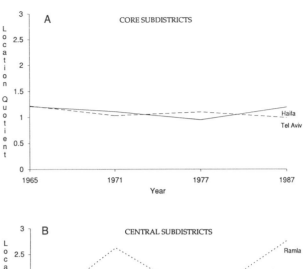

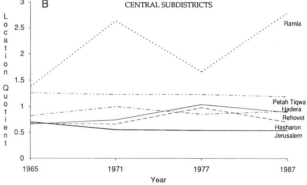

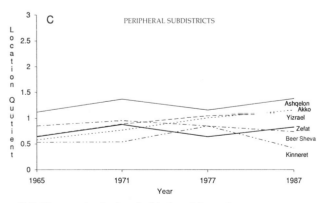

Figure 9.2 Changes in the level of industrial employment concentration by subdistricts, 1965–87

industrial dispersal policy. In the early stage, industrialisation was achieved by the introduction of large plants, most often textile. This phase was followed, however, by two processes working to increase levels of diversification: the opening of small non-textile operations and the closure, or reduction in size, of the early-phase large textile plants.

Analysis of levels of industrial diversification reached by 1987 reflects the dynamic nature of industrial development. A comparison of the rank order of each subdistrict on the measure of diversity along four points of time (see Figure 9.3) reveals the following observations:

1 In most areas, diversity levels tend to fluctuate through time, reflecting closures, layoffs, and opening of plants.
2 The long-term tendency in the core industrial regions is from high to medium diversity (see Figure 9.3a).
3 Of the six central subdistricts, two are the least diversified all or most of the time periods, two go from low to medium diversity, and only two are more diversified in 1987 than in 1965 (see Figure 9.3b).
4 All six peripheral subdistricts are ranked more diversified in 1987 than in 1965, though some of them went through substantial fluctuations (see Figure 9.3c).
5 Three of the peripheral subdistricts are ranked among the four most diversified areas.

These observations seem to lead to the conclusion that the regional industrial mix is in a constant state of flux. Most central regions and some of the peripheral regions show a long-term tendency towards greater specialisation. On the other hand, after the initial, single-branch, big push towards industrialisation that took place in the peripheral regions, there was room for certain processes to operate towards the reduction of over-specialisation. In some (semi)-peripheral areas, especially those adjacent to the central region, the long-term reduction of specialisation, coupled with government dispersal of incentives for industry and some local self-generated growth, acted to increase the levels of industrial diversity.

Another measure suggested by Gradus and Krakover (1977) was the index of concentration. This index, as measured by the location quotient of the proportion of employment in industry i in subdistrict j relative to its national proportion, is used to identify the degree of branch spatial concentration. It was found that the increase in the

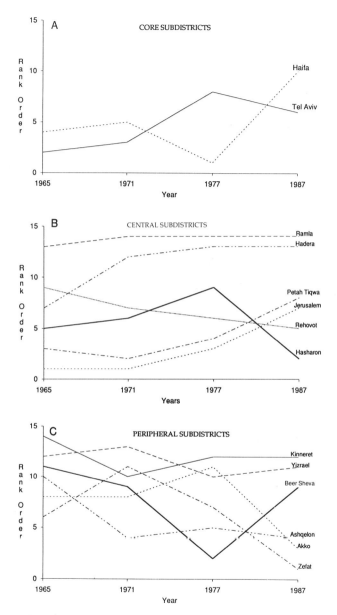

Figure 9.3 Changes in the rank order of subdistricts in terms of industrial employment diversification, 1965–87 (0 – most diversified; 15 – least diversified)

level of industrialisation in the 1960s in the peripheral subdistricts was achieved mainly through the introduction of laggard industries like textiles, apparel, wood products, and processing of non-metallic minerals. Two more advanced branches that were found concentrated in some of the peripheral areas were petrochemicals and machinery. On the other hand, the concentration index for the central subdistricts revealed specialisation in the growth industries of electronic equipment, aircraft industries, plastics, and metal products. The follow-up study by Gradus and Eini (1981) found slight changes to this pattern, generally in a positive direction. Their findings indicated that until 1978 the two growing industries of plastics and metal products achieved above-unity concentration in peripheral subdistricts.

Analysis of data for 1987 shows further penetration of some advanced branches of industry to the northern peripheral subdistricts. Thus, the plastics and metal products industries are represented in higher-than-unity concentration levels in all four subdistricts of the northern area, and machinery in three subdistricts out of the four. The situation in the southern subdistricts is less optimistic. Here, none of these branches have achieved high levels of concentration. Another advanced branch that has shown little concentration in any of the peripheral areas is electronic equipment. This branch's concentration is limited to the core area of Haifa and four of the six central subdistricts.

In sum, the three indices applied to evaluate the effectiveness of the dispersal policy seem to suggest that processes of industrialisation have continued to operate in the peripheral areas even after the initial big push towards industrialisation. These processes are constantly shaping and reshaping the regional industrial strucuture. However, it seems that there is enough evidence to support the hypothesis that the main thrust of the national industrial dispersal policy is felt primarily in the semi-peripheral regions adjacent to the central subdistricts.

10

STAGNATION OF THE INDUSTRIAL BASE IN THE PERIPHERY AND THE CASE FOR DIVERSIFICATION

The industrial plants established in Israel's peripheral regions during the late 1950s and early 1960s had a major role in creating a stable economic base in the development towns. However, during the 1970s, it was increasingly argued that this industrial base had become a burden to the development of these towns as a gradual rise in the level of education and skills has rendered these 'old generation' plants less suited to the employment skills of the local labour force (Kipnis 1976).

Many of the plants established in the more remote development towns acted according to the principle of 'maximising subsidies' – that is, attempting to derive as much financial support as possible from the government – while the owner's investment remained minimal. Consequently, the incentive to operate these plants efficiently was lacking, and the return on their investment was low (Shinan-Shamir 1984). Moreover, government incentives were especially attractive to inexperienced entrepreneurs and to firms with no alternative financial resources. Thus, many of these companies experienced a high turnover rate, with plants closing down only a few years after establishment.

In the development towns, single-plant firms had the highest closure propensities (See Table 10.1). Examination of plant closures during the first five years of existence revealed that the closure rates of young plants owned by non-local, single-plant firms was at least double that of young plants of other ownership types (See Table 10.1). Moreover, while locally owned plants had in most cases only limited amounts of fixed capital, plants owned by non-local single-plant firms were much larger and more capital-intensive, and were usually established with generous government assistance, which was the major motive for their location in development towns.

Table 10.1 Plant closures in Israel's development towns, in selected groups of plants by ownership type*

Ownership type	Plants existing in 1975		Plants existing in 1966		Plants established from 1966-78			Plants established from 1961-70 reaching age 5		
	Total	Closed until 1983 (%)	Total	Closed until 1983 (%)	Total	Closed at ages 0-5	(%)	Total	Closed at ages 6-13	(%)
Single-plant, local owners	34	35.3	12	41.7	71		16.9	36		33.3
Single-plant, non-local ownership	41	24.4	24	50.0	65		32.3	31		25.8
Single-plant, non-local head office	27	33.3	20	55.0	47		31.9	17		0.0
Multiplant, non-local head office	116	18.1	60	30.0	97		15.5	55		12.7
Muliplant, local head office	1	0.0	0	–	0		–	0		–
Rural-cooperative sector	25	4.0	26	7.7	10		0.0	12		0.0
Unknown	4	100.0	10	100.0	25		80.0	8		75.0
Total	248	23.0	152	38.2	315		26.3	159		20.8

Source: Razin and Shachar 1987.

Notes: * The first two groups include plants employing 20 or more workers in the base year 1966 or 1975. The other two groups include plants employing 20 or more workers for at least one year.

Hence their high closure propensity at a young age represented a considerable waste of public resources (Razin and Shachar 1987).

Closures were usually a consequence of the establishment of plants without prior examination of appropriate technologies and marketing possibilities. Plants also suffered from management failures, since the local population lacked employees with high management skills. Efforts to persuade managers and qualified workers to move to development towns, particularly by offering inexpensive housing, were unsuccessful largely because the development towns were unattractive to their families. The low-level education systems and the lack of job opportunities for professional wives were particular obstacles. In addition, those who were ready to commute from central regions to remote development towns were often inexperienced managers who considered their jobs in the remote plants as stepping-stones towards better positions in the metropolitan regions (Shinan-Shamir 1984).

While failing small plants had to close down, the larger, old generation plants survived by infusions of government funds in times of crisis. These plants were characterised by obsolete technologies, low-level work ethics, hidden unemployment, and constant profitability problems. Government policy prevented immediate unemployment problems, but in the long run it may have impeded the entrance of new industries and contributed to the negative image of development towns (Shinan-Shamir 1984).

These old generation plants offered routine, low-wage jobs and lacked almost any internal promotion ladders. Severe unemployment problems emerged, particularly among the second generation returning to their towns upon completion of compulsory army service. Some of these youngsters, in their early twenties, had acquired technical skills in the army, vocational schools, or general high schools. These qualifications, as well as exposure to big city lifestyles, had raised expectation levels. As a result, many were not prepared to work, as their parents had, in unskilled occupations for salaries that roughly equalled unemployment benefits. Frequently, they could not find jobs commensurate with their skills and expectations; consequently, those who could qualify for better employment elsewhere left the development towns. Others preferred to remain unemployed, and some local plants had increasingly to rely on mostly Arab workers, who had to commute long distances to work.

A particular problem has been that these new entrants to the labour market sometimes rejected jobs in solidly based new industries that offered average wages, an improved quality of working life, and opportunities for advancement, but which also required hard work and a high level of commitment. The complaints against work in these plants were that they presented high demands in terms of meeting production schedules, accuracy, and quality control. However, starting salaries in these plants were still unattractive when compared with unemployment benefits paid to discharged soldiers. It was thus argued that these new labour market entrants overlooked the benefits of the opportunities of being trained on the job where they could gradually achieve both new skills and increased pay (Shinan-Shamir 1984). This was partly due to the atmosphere prevailing in some development towns of general dependency on the government for employment provision. This dependence largely developed as an historic result of the way in which the new immigrants were absorbed into these towns by the public agencies during the 1950s and 1960s (Gur 1990).

Consequently, most towns continued to depend on public assistance for their demographic and economic development, and did not reach a 'takeoff' point characterised by independent economic growth. In many cases, massive aid for establishing new plants sufficed to attain only modest growth or to prevent a severe process of decline that would have occurred if market forces were left to act alone. As industrial growth in Israel slowed down and public funds available for industrial dispersal begun to contract, unemployment re-emerged as a problem of major significance. Rates of unemployment in development towns increased from around 5 per cent in 1970 to 11.5 per cent in 1986 – twice as much as the national average (Lavy 1988). The development towns were more sensitive than the central regions to cyclical economic fluctuations. Moreover, the rate of unemployment, particularly for males, increased much faster in the development towns during the 1980s than in the central regions.

Mineral-extracting plants in the Negev, primarily the Dead Sea Works, have become an exception to the gloomy picture of industrial development in Israel's southern periphery. After two decades of enormous investment and huge losses, the Dead Sea Works began to show a profit in 1971 and soon became one of the most profitable industrial enterprises in Israel. However, being extremely capital-intensive, the improved conditions of the mineral-extracting industry did not result in employment growth. In addition, profits were transferred to the government (Gradus 1984) and, hence, the main local multiplier effects were only via increased subcontracting activity and a local income multiplier.

Calls for a revision of existing priorities emerged during the late 1970s; it had been realised that job-creation in development towns might not lead to urban growth and alleviation of economic problems in the long run unless the new employment contributed to improved standards of living (Bar-El *et al.* 1982). Reduction of regional inequalities was emphasised, and policy evaluations pointed to the undiversified industrial structure of development towns and the domination of large capital-intensive plants in 'laggard' industries. The need for diversification of the manufacturing branch structure in the periphery, and for the dispersal of growing industries, particularly in high-technology sectors, has therefore been a major regional development theme that has dominated studies since the mid-1970s and throughout the 1980s (see Tables 2.1 and 3.1; Kipnis 1976; Gradus and Krakover 1977; Gradus and Eini 1981; Justman 1985; Weinblatt and Luski 1986; Shefer and Frenkel 1989).

However, these proposals, intended for implementation during the 1980s, were largely based on post-Six Day War realities, which were characterised by rapid industrial growth and structural change. By the late 1970s, these realities were changing. The economy became increasingly stagnant and public policy budgets began to contract. Israeli industry was also headed towards stagnation, with the exception of rapid growth in the electronics industry (see Table 2.2 and Figure 2.4). Growing industries, particularly electronics, however, were hardly an employment substitute for declining industries, such as textiles, in the periphery.

In addition, a new major spatial goal of wide-scale settlement in the occupied territories was pursued by the right-wing Likud government (see Table 2.1; Newman 1984), with direct effort having been made since 1981 to industrialise some of these settlements. Preferential assistance status (zone A or A+), similar to that of the peripheral development towns, has been granted to these new Jewish settlements, which are in many cases within commuting distance of the Tel Aviv or Jerusalem metropolitan regions.

Thus, efforts to diversify the manufacturing branch structure in development towns came too late. The opportunity to disperse the expanding defence industry was particularly missed. The defence industry grew immensely throughout the late 1960s and 1970s, but then faced tight economic conditions and contracted during the 1980s. The two large government-owned enterprises – Israel Aircraft Industries and Israel Military Industries – grew rapidly to employ altogether about 34,000 workers in 1979. However, after reaching a peak of about 37,000 workers in 1985, they began to contract and in 1989/90 employed about 27,000 (see Chapter 13). Calls for dispersing a substantial portion of the defence industrial complex to the south were frequently repeated throughout the 1980s, disregarding the changed circumstances under which such wide-scale dispersal of the defence industry was no longer feasible.

Acknowledging the low probability of success in attempts to disperse most advanced industries to remote development towns, Shinan-Shamir (1984) called for favouring plants offering a diversified range of jobs and internal promotion ladders, which are not necessarily limited to high-technology industries. Other plants that are limited to offering only low-skilled or highly professional jobs may not be able to provide employment appropriate for those segments of the labour force present in development towns. The significance of plant size composition has also been stressed. Bregman (1987) has argued that small plants are more efficient than large ones in the Israeli economy, since their decisions are less distorted by government intervention. Kipnis (1977) called for preference of medium-sized plants over large ones in the development towns, due to their stronger local linkages. Shinan-Shamir (1984), on the other hand, viewed product and technology as the relevant attributes influencing income levels and quality of working life, the size of plants being only a reflection of these two factors. All of these proposals were difficult to implement under conditions of slow industrial growth and meagre public resources, and had very little impact on policy formulations and practices.

Since the late 1970s, with the economic slow-down and the rise to power of the right-wing Likud government, the commitment, as well as the ability, of the government to continue supporting large problematic plants gradually declined. Some large old generation plants, mostly in the textile industry, which had survived for nearly twenty years through massive public support, were gradually abandoned. Some of them closed down, while others attained profitability through change in ownership, lay-offs, and restructuring, mainly diversifying into fashion and reducing the production of intermediate textile products. Thus, the role of these first-generation plants in the economic structure of the development towns has shrunk. Such a policy shift could have been interpreted as a first phase in efforts to restructure the industrial base of development towns. However, new plants replacing old generation industries were not necessarily in those technologically advanced areas capable of pulling the development towns into a long-term growth process. Many of them, in fact, possessed the same characteristics as earlier industries that had subsequently closed down (Schwartz 1986).

11

INERTIA AND INCREMENTAL CHANGE

An evaluation of the incentives for industrial dispersal

EVOLUTION, ATTRIBUTES, AND DISTORTIONS OF POLICY MEANS

The major incentives for industrial dispersal were incorporated in the Law for Encouraging Capital Investment. This law was originally passed in 1950 with the purpose of encouraging capital investment, particularly foreign investment, in the Israeli economy. The benefits provided by the law included tax concessions, reduced obstacles to importing inputs, and permission to transfer profits abroad. The law was amended twice during the 1950s with the intention of strengthening and extending the incentives by subsidising capital investment. Population dispersal has been stated as one of the goals of the law since 1950, but only in 1959 were specific measures, differentiating between central areas and development zones, incorporated in the law. The public organisations in charge of implementing the law – The Investment Centre and the Industrial Development Bank of Israel, as well as the Development Regions Unit in the Ministry of Commerce and Industry – were all established during the 1950s. Thus, the basic system of incentives for industrialisation and industrial dispersal, which prevails to the present, was formulated during that decade.

The Law for Encouraging Capital Investment defined criteria for granting an 'approved enterprise' status to investment plans in manufacturing and tourism. In early periods there was an emphasis on import substitution and on securing the availability of products regarded as essential for Israel's local markets. However, later the criteria mostly referred to the location of the investment and to its contribution to exports (Bar 1990). Export requirements were raised, particularly for plants located in central regions. At a certain phase, specific industries, not regarded as particularly beneficial to the Israeli economy, were included in a blacklist of industries that were ineligible for 'approved enterprise' status. Exceptions to this rule were occasionally made in the cases of plants wishing to locate in the most depressed development towns.

The incentives granted to 'approved enterprises' included mainly grants and subsidised loans, as a proportion of the approved capital investment, and tax exemptions. The specific terms of incentives changed frequently, but always depended on the location of the investment. Additional benefits granted to plants located in the periphery included subsidised land and infrastructure developed by the government-owned Industrial Building Corporation. Plants located in development zones also occasionally received preference in obtaining government contracts, such as those of the Ministry of Defence. Nevertheless, it is uncertain whether this preference was sufficient to compensate for the distance from major defence enterprises which offered most subcontracts.

Export and R&D subsidies have been the two major industrial subsidies other than those incorporated in the Law for Encouraging Capital Investment. Export subsidies have existed since the 1950s, and, in the form of exchange rate insurance, were still the most costly industrial subsidies in 1990, in spite of frequent attempts to reduce and gradually eliminate them. The incentives of the Law for Encouraging Capital Investment were second and, since the 1970s, support for industrial R&D was third in terms of government expenditure (Rivlin 1991). Export subsidies have been non-spatial, whereas the Law for the Encouragement of Industrial Research and Development, in force since 1985, gave a slight advantage to R&D projects in locations benefiting from the most preferred 'A' development zone status, which entitled them to a grant covering 60 per cent of R&D expenses rather than the 50 per cent offered in other regions. However, this had only a marginal impact on location decisions, particularly as 'start-up' companies, embarking upon their first R&D programme, could obtain a 66 per cent grant up to a certain limit everywhere.

The extensive incentive system which accompanied the rapid industrialisation of the late 1950s and early 1960s led to the shift in Israel's industrial geography. Governmental development loans accounted for as much as 40 per cent of the total investment in manufacturing in Israel during the second half of the 1950s. During the second half of the 1970s, the share of the government's assistance reached a new high, due to the accumulation of various modifications which improved the terms of incentives, and to rapid inflation which reduced debt payments in real terms. In 1981, subsidies and grants to the Israeli business sector reached a high of 14.6 per cent of the GNP (Rivlin 1991). However, for the remainder of the decade, the incentives were adjusted to inflation, becoming smaller and less attractive as pressures on the government's budget accumulated. Hence, by 1989, subsidies and grants to the business sector amounted to only 4.9 per cent of the GNP.

The incentives produced certain distortions in the patterns of investment which became more and more evident as the system was continued over decades. These distortions created particular problems for the industrial development of peripheral development towns.

The most evident distortion caused by the incentives of the Law for Encouraging Capital Investment lies with the establishment of capital-intensive plants in peripheral areas which had been granted development zone status mostly in order to alleviate unemployment problems (Zilberberg 1973). The high capital intensity in the periphery was more evident in the past, when the incentives were more effective; nevertheless, it still existed in the 1980s, even when interregional variations in the manufacturing branch composition, and particularly the dominance of the capital-intensive mineral-extracting industries in the periphery, were taken into account (see Table 11.1). Thus, plants receiving substantial assistance created few jobs, and usually only low local multiplier effects to compensate for their high capital intensity (Razin 1988a).

A second distortion was that plants enjoying the highest effective incentive to locate in the periphery were those engaged mainly with production of mature products for the Israeli market, and which had poor prospects of profitability. Development towns attracted non-exporters, since plants exporting a large proportion of their output could also get most incentives in central locations. The gap between export requirements from plants in development zones and in central areas was widest in traditional non-growing industries, considered as low priority industries from the point of view of national economic growth. Investment in such traditional industries was particularly attracted to development towns, since the government was ready to support them only if they exported a very high proportion of their output, or if they contributed to the alleviation of unemployment problems in the most deprived development towns (Schwartz 1986).

Table 11.1 Israeli industry – indices of capital intensity by district and selected branches, 1982 (national average for each branch = 100)

District	Industry total		Chemical products		Textiles	
	Capital-ouput ratio	Capital per worker	Capital-output ratio	Capital per worker	Capital-output ratio	Capital per worker
Tel Aviv	67	60	79	41	113	93
Central	83	80	69	48	95	104
Haifa	105	133	113	127	52	52
Jerusalem	91	74	73	41	–	–
Northern	104	108	73	104	129	127
Southern	138	189	101	135	91	104

Source: Borukhov 1989, based on Central Bureau of Statistics, *Survey of Capital Stock in Industry* 1982.

The behaviour of a firm engaged in the production of agricultural machinery, located in Israel's southern coastal plain, presents a typical example of the manipulation of the incentive system. The firm operated a plant in a location which was not entitled to development zone incentives. When it began to expand and export during the late 1970s, it sought 'approved enterprise' status. However, the proportion of exports from its total sales did not fulfil the requirements for obtaining such a status in a central location. Thus, the firm proposed establishing a new plant in an 'A' development zone. Production for the Israeli market was transferred to the new plant, which enjoyed the preferred incentives. Export production stayed in the old site which could now apply for an 'approved enterprise' status, having a very high export rate.

The incentives also had a weak influence on the location of R&D-intensive facilities because the generally non-spatial R&D subsidies have usually been far more significant for such functions than the capital subsidies. Since 1988, R&D of military products for export has been entitled to a significant advantage in an 'A' development zone – a grant of 45 per cent versus a grant of only 30 per cent for such products in central regions. However, this spatial incentive has been implemented just as public policy and economic circumstances have been shifting the focus of industrial R&D from military to civilian applications.

Tax concessions caused another distortion. The effective incentive to disperse has been higher for projects aimed at giving short-term returns, such as low-risk traditional production lines, rather than for long-range investments, which include new high-risk products and processes (Schwartz 1985). On the same grounds, the incentives attracted low-risk branch plants that were established in order to expand a firm's production capacity, rather than plants engaged with new products.

Finally, incentives attracted to development regions those enterprises and projects with a weak financial base which were most dependent on public financing (Schwartz 1986). Establishing plants in development towns and utilising government incentives was especially attractive to small, non-locally-owned, single-plant firms. These types of firms, while having the ability to negotiate government incentives and establish plants with substantial capital investment, frequently lacked credibility and a sound financial base. Screening by public authorities was not always effective, particularly when the proposed investment was in the most depressed towns. The Investment Centre was frequently criticised for approving investments even when feasibility studies led to a negative recommendation (State Comptroller of Israel 1991). Government incentives therefore attracted to development towns many plants that were owned by inexperienced entrepreneurs and firms having poor previous records (Razin 1988b; Schwartz 1985, 1986).

An analysis for the period 1977–84 demonstrates the tendency of the law to attract traditional non-growth industries to development towns (Schwartz 1989). An examination of the industrial development in Ofaqim (Schwartz 1986), one of the more problematic development towns in the south which enjoys an 'A' development zone status, showed that government incentives increased the profitability of investment for exactly those types of firms which would have already tended to be attracted to development towns. Such investments were made in plants having a high possibility of encountering difficulties. Indeed, of the new plants which obtained 'approved enterprise' status in Ofaqim during the period 1977–83, 27 per cent were not established at all, 18 per cent were closed down soon after establishment, and the remainder were investing at a much slower pace than initially planned. A frequent pattern was that plants which received large public aid for financing their investment closed down within a few years due to lack of working capital, while the government again supported the establishment of new plants with very similar attributes.

THE NEED FOR REVISION

Governmental policy responded to external pressures and the increasing difficulties of development towns only by incremental modifications in the terms of incentives. Subsidised loans were revoked and exchanged for increased grants that were less affected by inflation levels. Export requirements for receiving 'approved enterprise' status were also raised in the periphery. However, since the late 1980s, the free trade agreement with the United States restricted the ability to explicitly use exports as a criterion for granting incentives. An option to exchange grants for extended tax exemptions has been offered since 1985 with the intention of attracting more profitable R&D and skill-intensive industries to development towns (Ministry of Industry and Trade 1986). This option attracted mainly those firms establishing their plants outside a development zone.

The numerous incremental modifications made in the spatial policy have not touched its fundamentals, nor succeeded in significantly updating criteria and measures which have been based on the realities of the 1950s and 1960s. Incentives have been adjusted to inflation levels and to the reduced ability of the government to finance expensive policy measures, but clearly have fallen short of solving long-term profitability problems in the periphery and distortions caused by the policy measures themselves. Major failures have been the inability of the government to: (a) react to the clear need to revise the map of development zones; (b) reorient measures towards new types of enterprises; (c) decentralise agencies engaged in development efforts; and (d) re-evaluate general priorities in public efforts to promote the periphery.

Revision to the map of development zones

The map of development zones evolved during the late 1950s and early 1960s. Lists of development towns were prepared and constantly modified. Development areas included the peripheral regions in the north and the south, the Jerusalem area, and towns populated mainly by new immigrants in the central coastal plain. In 1967, a map of development zones was incorporated for the first time as an integral part of the Law for Encouraging Capital Investment (see Figure 11.1a). This map has been revised only once, in 1972 – a revision which slightly reduced the spatial extent of

INDUSTRIAL GEOGRAPHY IN A PERIOD OF ECONOMIC STAGNATION

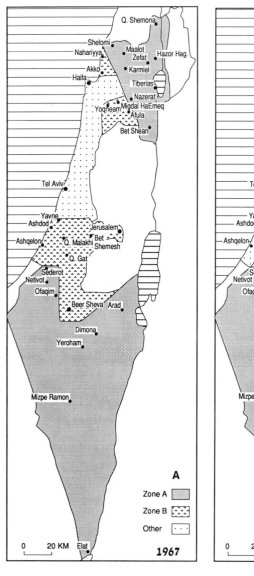
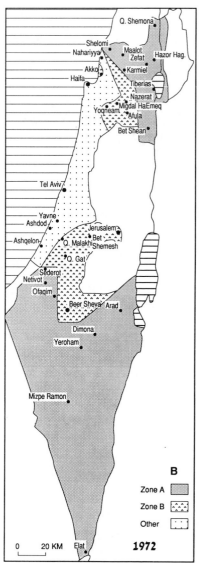

INERTIA AND INCREMENTAL CHANGE

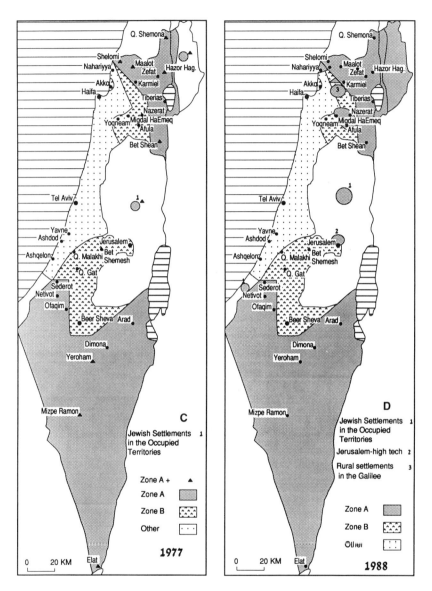

Figure 11.1 Israeli development zones according to the Law for Encouraging Capital Investment (including temporary changes), 1967, 1972, 1977, 1988, 1991

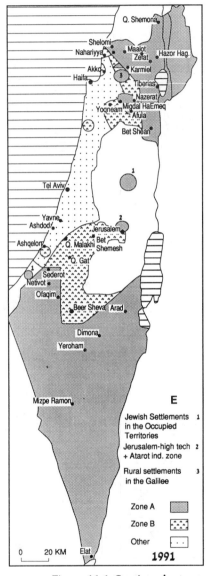

Figure 11.1 Continued

development zones and eliminated the preferred 'B' development zone status of towns in Israel's northern and southern coastal plain. Parts of the western Galilee were also downgraded from an 'A' to a 'B' development zone (see Figure 11.1b).

The outdated 1972 statutory map of development zones has remained unchanged to the present (1992). On this map, layers of temporary changes were made, mostly undertaken to solve short-term problems in specific localities, or as the result of pressure from influential mayors and industrialists. In 1975, eight towns which were then considered to be most problematic received an 'A+' status (see Figure 11.1c). In addition, Jewish settlements in the occupied territories received an 'A' or 'A+' status. High technology enterprises in Jerusalem also received an 'A' development zone status.

Significant temporary changes have been made since 1978, particularly in the Galilee area (see Figure 11.1d). Three influential mayors were able to upgrade the status of their towns (Migdal HaEmeq, Nazerat, and Afula) from 'B' to 'A' development zones. Nahariyya's status was upgraded shortly before the 1982 Lebanon war, when the town was shelled from Lebanese territory and suffered from terrorist attacks. New Jewish rural settlements established since the late 1970s in an effort to secure Jewish presence in predominantly Arab areas in the Galilee area also received an upgraded status, as did a limited number of Arab settlements. Thus, areas enjoying a development zone status gradually began to approach the vicinity of the Haifa metropolis. Further incursions of the development zones into the central regions occurred at the onset of the 1990s (see Figure 11.1e). This process aggravated distortions in which towns having no preferred status faced competition from nearby towns enjoying an 'A' development zone status. In addition, remote towns had to compete with towns in the vicinity of metropolitan areas which offered potential investors the same 'A' status incentives.

Temporary changes were made without considering their wider implications for other towns and regions, and all changes extended the map of assistance. Whereas the basic statutory map defined regions entitled to development zone benefits, the temporary changes were all point-oriented; that is, directed towards specific localities. This reflects the basic difference between policy measures resulting from a comprehensive planning approach, and those produced through political pressure leading to incremental modifications.

The increasing number of Arabs employed in 'old generation' plants in development towns and the suburbanisation of high-technology branch plants to those development towns at the fringes of the metropolitan areas aggravated the problem of plants receiving government incentives for the purpose of solving employment problems in the development towns while employing a high percentage of non-locals. Only 37 per cent of the employees in Migdal HaEmeq's industry, for example, were local residents in 1984, while most highly paid skilled employees commuted from higher status localities in the Haifa metropolitan area, and low-paid unskilled labour came from nearby Arab towns (Razin 1986).

During the 1980s, various government agencies and committees proposed new maps of development zones, mostly reducing the extent of the assisted areas and removing those places in the vicinity of the metropolitan centres (Efrat 1987). These proposals were based on criteria such as distance from central areas and local demographic, economic, and social conditions. However, successive government efforts at implementation have been blocked by local political interests. This failure can be traced to political circumstances in which influential leaders representing development towns near the metropolitan areas have successfully defeated any move to downgrade their assisted status. The fate of the Law of Development Towns and Regions demonstrates this point. Passed in the Knesset (Israel's parliament) in 1988, the law gave development towns extensive benefits in the form of enlarged governmental financial transfers to local authorities and various kinds of financial benefits to their residents. Implementation of the law, which passed despite the opposition of the Ministry of Finance, was linked to a reclassification of development towns, removing some of the closer towns from the list. As a result, a lobby of mayors from these adversely affected towns agreed to support the delay of the implementation of the law in return for the abolition of all proposals for reclassification. In addition, it became more difficult to justify a reduction in the geographical extent of the development zones in the North and the South without reducing the extent of the development zone in the occupied territories, a step which would have faced strong political opposition from the right-wing ruling coalition.

The prime losers from the delays in revising the map of development zones have been remote development towns, unable to compete with those places closer to the fringes of the expanding metropolitan areas. It can be argued that while several towns, located within commuting distance of metropolitan areas, still suffer from social and economic problems, remedies may not require massive government support for further industrialisation, but lie more in the area of social and housing policies. Thus, there remains a need to overcome political obstacles in order to radically alter the map of the development zones. Such a revision should smooth sharp zone boundaries so that incentives would increase more regularly with distance from the metropolis, and differences between adjacent towns would be small.

Reorientation towards new types of enterprises

The ineffectiveness of incentives for industrial dispersal is also due to the fact that they are oriented towards those industries that no longer contribute to the economic growth of development towns. As stated at the beginning of this chapter, the incentive system has been unable to influence substantially the locational behaviour of advanced industrial activities. The continuing emphasis on subsidising capital stock stands in contrast to the nearly total absence of any subsidies to labour or to working capital, and this has reduced the long-term impact of the policy (Schwartz 1989). Reorientation of public assistance towards local entrepreneurs and establishment of new firms has consequently received growing attention during the late 1980s (see Chapter 15). However, the bulk of public assistance to economic development in general, and to the economic development of peripheral regions in particular, has remained within the traditional framework of the Law for Encouraging Capital Investment.

Decentralisation of agencies engaged in development efforts

Another needed reform has been to decentralise those agencies in charge of the incentive system. The organisational structure in charge of implementing the industrialisation policy was formed during the 1950s, mostly within the Ministry of Industry and Trade. The 1950s were characterised by a young and dominant central government, which demonstrated coherent and flexible action and had almost absolute control of large-scale capital investment in the Israeli economy. During the 1980s, while the organisational framework remained basically intact, central government demonstrated much weaker and fragmented action and faced difficulties in coping with both external pressure groups and internal conflicts of interests among Ministries and other public agencies. The government also faced severe financial constraints and had little flexibility in redefining priorities in the national budget. Hence, there does seem to be some justification for decentralising some of the responsibility for managing the public industrial development efforts to local authorities (see Chapter 14).

Assigning the private sector a greater share of responsibility for implementing policy could be another option for organisational reform. This could be done either by contracting with banks and other firms to implement more efficiently and effectively some phases of the public policy, or by establishing new organisations in the form of public–private partnerships for promoting industrial development (Felsenstein *et al.* 1991).

Re-evaluation of priorities

Lastly, re-evaluation of general priorities in spending money to promote Israel's periphery has also been long overdue. Tourism development has been perceived as complementary to industrial development, but reference to other sectors of the economy has been minimal. Schwartz and Felsenstein (1988) examined the option of facilitating producer services, particularly business services, in the periphery. They found a very limited potential for such action due to the relatively small distances in Israel, which make it unlikely for high-level producer services in the periphery to compete effectively with such establishments in the Tel Aviv and Haifa metropolitan areas.

Rivlin (1991) called for the abolition of nearly all industrial subsidies in favour of a general lower rate of corporate tax. Government intervention, according to this argument, should limit itself to its traditional role in developing infrastructure. A much publicised proposal has been to substitute massive investment in transportation (Economic Models 1989) and communications (Salomon and Razin 1988) infrastructure for the capital incentives for industrialisation in the periphery, which have proved to be expensive and ineffective in job generation. The major component in this alternative has been to develop a network of freeways that will reduce all parts of Israel, from Qiryat Shemona in the north to Dimona in the south, into three large and closely linked metropolitan labour markets. The freeway network would lead to the dispersal of industrial and other economic activity throughout the coastal plain. Population would thus disperse to more distant localities, enjoying the convenience of being able to commute to one of three metropolitan centres: Haifa, Tel Aviv, or Beer Sheva. Thus, the link between industrial investment and spatial policy would not be necessary any more, and government resources would achieve far more effective results (Economic Models 1989).

Some fears have been expressed that such improvements in the transportation system could also have an opposite effect, enabling greater spatial specialisation in which top-level economic functions and the affluent population would become even more concentrated in the better-endowed central regions. Nevertheless, the lack of serious attempts to divert resources from large-scale capital subsidies to transportation and communication infrastructure has nothing to do with this last argument. Rather, it has been a consequence of the lack of any political or bureaucratic forum having sufficient authority to make such a major redistribution of resources among government ministries and agencies.

Two decades of inertia and small incremental changes in governmental policy ended abruptly in 1990, when immense pressures created by the wave of mass immigration from the USSR created a new readiness to reinforce existing incentives and accept new ideas for job-creating policy tools (see Chapter 15). These pressures have led to the latest of many amendments in the Law for Encouraging Capital Investment, made in 1990, which added an option of government guaranteed loans taken out by investors for the purpose of economic development. The guarantees generally cover up to two-thirds of the proposed investment of approved enterprises (Ministry of Industry and Trade 1991). Terms have been somewhat more favourable in development zones, but in general this channel has been most attractive to plants in central regions which were not entitled to grants. Early implementation of this new funding channel does indeed indicate that it has increased the attractiveness of central regions. The government guarantee scheme may be most attractive for medium-sized firms and for high-risk investments, whereas extended tax exemptions may still remain most attractive for large established enterprises. In both cases, the substitution of grants with other incentives which are less contingent on location may increase the attractiveness of central regions, after a prolonged period in which the level of support in these regions was eroded. The new emphasis on encouraging small businesses and the establishment of technological incubators may also reduce the significance of the spatial element in the public economic development policies (see Chapter 15).

12

CORPORATE GEOGRAPHY AND THE CRISIS IN THE FEDERATION OF LABOUR ENTERPRISES

The Israeli industrial sector is characterised by relatively small enterprises. Its modern roots were largely sown by petty entrepreneurs from Europe, who came to a small and isolated country. Local market size has constrained the growth of Israeli enterprises to modest proportions by global standards. The Arab boycott has discouraged investment of large multinationals, with a few exceptions which are mainly in the electronics industry in which Israel enjoys a comparative advantage. In addition, Israel's geopolitical position has not made it easy for local firms to penetrate global markets to a great degree.

After Israel's independence, a process of organisational concentration in industry was triggered by initiatives of large-scale, publicly owned enterprises, first by those owned by Israel's General Federation of Labour, and later by those owned by the government. As the Israeli economy matured, some private firms grew to dominate local markets in various manufacturing branches. However, these could be regarded, at most, as medium-sized enterprises by global standards. In private enterprises, family ownership has still been most common. Only in a few cases are corporate shares split among several shareholders who do not constitute inter-family partnerships.

From a spatial point of view, it was the public policy of industrial dispersal which created a large-scale branch plant economy in the development towns and produced a clear hierarchical spatial division of labour. Industrialisation of the development towns was not indigenous, but to a large extent depended on external investment attracted by public support.

LOCATION OF HEADQUARTERS OF THE LARGE INDUSTRIAL FIRMS

The head offices of the large Israeli industrial firms are generally concentrated in the central regions of the country. The largest concentration is in the Tel Aviv metropolitan area, and a smaller concentration occurs in metropolitan Haifa (see Table 12.1). Only a few head offices of large firms are located in development towns. The concentration of head offices in Tel Aviv is perhaps not as striking as could have been expected. This is mainly due to the relatively modest size of most of Israel's industrial enterprises. Nearly 80 per cent of the head offices of the 124 large industrial firms in Israel were adjacent to their main plant in 1981 (Razin and Shachar 1990). However, when operating a separate head office, industrial firms preferred in nearly all cases to locate it in the city of Tel Aviv. Moreover, in many cases, head offices located in the Tel Aviv metropolitan area managed plants that were established in development towns all over the country, while head offices located in other cities tended to manage industrial activity mostly in their hinterlands.

Head offices might be established close to the main plant for historical reasons or for easy communication with the largest production unit. Head offices located at a distance from the main plant were usually part of a multiplant structure with production units scattered over the country, particularly if their main plant was situated in a development town, or those which, in addition to industry, were also active in tertiary activities.

Some of the large industrial firms were actually subsidiaries of holding or investment companies; hence, a distinction needs to be made between the head office of the industrial firm and the headquarters of the parent company, defined as holding at least 50 per cent of the shares of the firm in question. (In the few cases of a foreign parent company, the location of its subsidiary's head office in Israel is used for the analysis.) The concentration of these top control units in the Tel Aviv metropolitan area, particularly in the city of Tel Aviv, was much more striking than that of direct head offices of industrial firms (Razin and Shachar 1990). Haifa, on the other hand, was a much smaller centre of top control units.

The development towns are the heart of Israel's externally controlled economy. Beer Sheva is the largest development town and the only one where substantial employment can be found in

Table 12.1 Location of head offices of the 20, 100 and 150 largest industrial firms in Israel, 1981–8*

Location	20 largest			100 largest			150 largest	
	1981	1982	1988	1981	1982	1988	1982	1988
Tel Aviv	4	5	5	29	28	21	38	33
Tel Aviv suburbs	5	4	6	20	19	28	27	34
Haifa Metro	4	4	5	13	17	14	21	17
Other towns in the Coastal Plain	3	2	1	18	18	16	26	23
Rural settlements in the Coastal Plain	0	0	0	5	6	7	13	15
Total – central regions	16	15	17	85	88	86	125	122
Jerusalem	1	1	1	3	3	4	4	5
Beer Sheva	2	2	1	2	2	2	2	2
Other development towns	1	2	1	6	4	4	5	8
Rural settlements in development regions	0	0	0	4	3	4	14	13
Total – develop. regions	4	5	3	15	12	14	25	28
Total	20	20	20	100	100	100	150	150

Source: Dun & Bradstreet International, *Dun's 100, Israel's Leading Enterprises*, various years.

Note: * The list only includes operating companies (that is, excludes holding or investment companies) in which over 60 per cent of volume of sales is derived from manufacturing. The companies were ranked according to their sales volume.

head offices of large industrial firms. Beer Sheva, however, is still far from constituting a control centre comparable to those in the central regions.

A striking contrast to the externally controlled economy in the urban periphery is the significant number of head offices of large industrial firms found in peripheral rural settlements (see Table 12.1). These relate mostly to industries owned by the kibbutzim (communal settlements), and their ultimate control has remained in the kibbutzim owning them. The rural-cooperative sector in Israel possesses an autonomous economic and social system, as well as local government and services. The cooperative structure of the rural sector has been used to some extent to overcome difficulties encountered by remote location (see Chapters 16–19).

THE SPATIAL ORGANISATION OF THE LARGEST MULTIPLANT FIRMS

An analysis of the spatial organisation of Israel's largest industrial corporations sheds light on the hierarchical spatial division of labour, characterising, in part, Israel's industrial geography. Razin and Shachar (1990), using data for the period 1979–81, looked at the spatial structure of the six largest firms in Israel in terms of sales volume: (1) Israel Aircraft Industries, the largest industrial government-owned concern, employing 20,000 people in 1980; (2) Israel Military Industries, which was a subsidiary of the Defence Ministry until 1990, with 14,000 employees in 1980; (3) Israel Chemicals, a holding company for the mining and inorganic chemicals firms owned by the government, with 5,600 employees in 1980; (4) Koor Industries, the largest industrial corporation in Israel, owned by the Histadrut (Israel's General Federation of Labour), with some 30,000 employees in 1980; (5) Clal Industries, a subsidiary of Clal (Israel) Holding Company, the majority of whose voting shares is held by two of the largest banking groups in Israel, with some 10,000 employees in 1980; and (6) Discount Bank Investment Corporation, an industrial holding company owned by IDB Bank-Holding Corporation, employing only 2,000 people in 1980 and 5,200 more in subsidiaries, where the company holds a minority of 20 per cent of the shares or more.

The headquarters of all of these firms are located in the Tel Aviv metropolitan area, four of them in the city of Tel Aviv. Israel Chemicals and, to some extent, Koor Industries are the only firms where a large part of the subsidiary and divisional head offices are located outside the Tel Aviv metropolis, mostly in the Haifa and Beer Sheva regions.

Of the large firms, employment was most dispersed outside of the central regions in Israel Chemicals, while the lowest dispersion level was that of Israel Aircraft Industries. Most of the production of the government-owned Israel Chemicals is oriented towards raw materials and was naturally located in the development regions of southern Israel. The employees of the two government enterprises, which mainly produce military equipment, were concentrated in the central regions. About 60 per cent of the employees of Israel Aircraft Industries were working in plants located near Ben-Gurion Airport in the outskirts of the Tel Aviv metropolitan area, and only 7 per cent were in development regions (see Table 12.2). Israel Military Industries has been establishing plants in development regions since 1967, but its activities have remained primarily concentrated in the central regions.

It would apparently be reasonable to assume that government-owned companies would be more responsive to the goals of the government spatial policy, but, at least for the firms analysed, this was not the case. Israel Military Industries and Israel Aircraft Industries were not very responsive to the government's capital incentives, which were mainly directed towards the private sector facing the problem of raising capital for large investments. Plans to relocate large portions of military industries to the sparsely populated south, where large tracts of undeveloped land isolated from residential areas are available, have existed since the 1960s and have recently been revived. Such relocation could reduce the risks associated with the production of armaments near populated areas, and vacate more valuable land for alternative uses. However, these plans were not implemented, partly because of the high cost of such relocation, and probably also because of the unenthusiastic attitude of management and employees towards moving southward from the Tel Aviv area to the arid peripheral region of the Negev.

Koor Industries is the largest and most diversified industrial firm in Israel. Koor is operated like all other business enterprises, but, being owned by the Labour Federation, it has been officially committed to goals other than profit maximisation (Barkai 1983), one of these being the industrialisation of development towns. A fast increase in the percentage of Koor's workers employed in development regions occurred between 1954 and 1969, and a slight

Table 12.2 Employees of large industrial firms located in development regions, by manufacturing branch (percentage)

Manufacturing branch	Employed in development regions[a]								Total Israeli industry 1978
	Aircraft Industries 1979	Israel Chemicals 1980	Koor Industries 1954	Koor Industries 1969	Koor Industries 1980	Clal Industries 1980	Discount Investment 1980		
Food	–	–	–	–	16.8	X	–		26.0
Textiles, clothing and leather	–	–	X	–	37.5	87.8	X		28.9
Mining	–	100.0	–	–	–	–	–		71.1
Non-metalic mineral products	–	–	6.8	21.2	42.3	21.1	–		33.9
Chemical, rubber and plastic prod.	X	37.8	X	78.2	43.0	X	100.0		34.7
Basic metal	–	–	47.5	53.6	45.7	27.8	0.0		41.9
Metal products, machinary and transport equip.	10.1	–	0.0	46.9	51.8	21.3	6.1		21.6
Electrical and electronic equip.	1.1	X	X	3.6	3.0	0.0	X		11.3
Total[b]	7.3	78.0	12.4	30.0	27.7	46.1	25.8		25.9

Source: Razin and Shachar 1990.

Key: – The firm has no activity in the branch. X The firm has less than 300 employees in the branch.

Notes: [a] Percentage employed in the development regions of the total number of employees in the firm occupied in the specific branch.
Development regions – 1972 definition (see Figure 11.1b).
[b] Including other branches not shown in this table (wood, paper, diamonds, printing, etc.).

decrease, from 30 per cent to 27.7 per cent, between 1969 and 1980 (see Table 12.2). The decrease in the percentage of total employment in the Koor plants in development regions between 1969 and 1980 appears to be mostly the result of the corporation's general shift to the production of electrical and electronic equipment, which has been concentrated in central regions. During the same period, employment in electronics grew from 28 per cent to 34 per cent of Koor's total employment, most of it in Tadiran (Israel's third largest high-technology military products manufacturer). The percentage of employment in the more dispersed branches declined over the period 1969–80: from 19.5 per cent to 11.7 per cent in non-metallic mineral products; from 20 per cent to 7.8 per cent in basic metal; and from 20.8 per cent to 14.7 per cent in metal products. Overall, the level of spatial dispersion in most branches rose during this period, but total dispersion remained stagnant owing to the changing industrial composition of Koor Industries' activities.

Employment in Clal Industries was more dispersed than at Koor (see Table 12.2). However, a high dispersion level existed only in the textiles, clothing and leather, where most of Clal Industry's employment – 40.3 per cent of the total – was concentrated in three large plants located in development regions. The other branches of Clal Industries were mostly concentrated in central regions. Hence, the tendency of the largest private and government-owned enterprises to disperse employment to the development regions was found to be lower than average when differences in the manufacturing branch structure were taken into account.

THE ROLE OF OWNERSHIP CHARACTERISTICS IN INDUSTRIAL DEVELOPMENT IN THE PERIPHERY

Industrial organisation, as manifested in the types of ownership of industrial plants, has had a major influence on the industrialisation process in Israel's peripheral areas (Razin 1988b). Firms of different types have reacted differently to the government inducements to locate in development towns.

Locally owned plants

Residents of development towns have benefited little from government capital incentives as they have lacked the capabilities to exploit this assistance. Locally owned plants consume more inputs locally than other plants, but their dependence on local and regional markets, which restricts their growth prospects, is even greater than their dependence on local suppliers.

Non-local single-plant firms

This type of ownership is unique to the externally controlled economy of Israel's development towns. Small entrepreneurs usually tend to establish their businesses in their towns of residence (Hamilton 1974). A decision of a resident from a central region to establish a plant in a remote development town can generally be explained by government incentives. The relatively short distances between most development towns and the central metropolitan centres made possible the widespread phenomenon of non-local single-plant firms.

The preference to retain residence in the centre of the country has been justified by the advantages of proximity to the centres of economic activity and by the involvement of some owners in more than one business. Moreover, owner-managers are attracted to stay in large urban centres due to educational and employment opportunities for children and spouse, as well as for the social and cultural amenities that are available for the entire family. Decisions of non-locally-owned, small, single-plant firms were more affected by the governmental incentive system than those of firms of any other ownership type. These small firms faced much greater difficulties than large multiplant ones in financing the capital investments; therefore, governmental incentives led to their high propensity to locate plants in development towns.

Externally owned small, single-plant firms present a particular problem for the development towns, as they are generally unstable (see Table 10.1) and form very few local linkages. Non-local, single-plant firms tend, more than others, to close their plants soon after opening them, and these types of firms often fail because they frequently lack a sound financial base.

Multiplant firms

Large multiplant firms take governmental incentives into their cost considerations, but in view of their general financial independence, they use the incentives in a more selective way. Hence, they tend to locate capital-intensive and non-exporting plants in the development towns. These plants are generally stable, but their geographical specialisation results in their offering only a narrower range of jobs.

The rural-cooperative sector

The location of plants owned by the rural-cooperative sector is governed by the location of the rural settlements themselves. Unlike local entrepreneurs in the development towns, the rural-cooperative settlements have the professional capabilities and the necessary business and political contacts to fully exploit governmental incentives for establishing highly mechanised and capital-intensive plants.

Plants owned by the rural-cooperative sector showed a striking stability until the early 1980s (see Table 10.1), a feature that could have resulted from the nature of decision-making in the kibbutzim, which puts greater stress on long-term policy. Nevertheless, the disadvantage inherent in these plants is that most of them only offer unskilled jobs to the residents of nearby development towns, while all high-skilled and top administrative positions are reserved for kibbutz members. Thus a social division of labour is constituted.

THE CRISIS IN THE FEDERATION OF LABOUR ENTERPRISES

The latter part of the 1980s witnessed the weakening of two of the major organisations that traditionally contributed to the industrialisation of Israel's periphery – the Federation of Labour and the rural-cooperative sector. The Israeli economy can be split into three major sectors of ownership: private, Federation of Labour, and public (see Table 12.3). The phenomenon of a significant portion of the Israeli industry, as well as a large bank and a major construction and development corporation, being owned by a national labour organisation is a feature unique to Israel. It reflects the major role of the labour movement in the nation-building process between the 1920s and the 1950s.

Koor Industries has been the flagship of the Federation of Labour industries, as well as of Israeli industry, for many years. Its rapid growth during the 1950s, as well as the slow-down in the 1960s, were discussed earlier (see Chapters 5 and 6). Rapid growth resumed in Koor during the 1970s, and the number of employees in the concern tripled from about 10,000 in 1969 to over 30,000 during

Table 12.3 Israel – employed persons in industry by sector of ownership, 1965–90 (percentage)

	Private	Federation of Labour [a]	Public [b]
Plants with 5+ salaried workers			
1965	71.6	16.7	11.7
1970	73.5	15.1	11.5
1980	62.0	20.3	17.7
1987	65.3	19.4	15.3
1988	66.6	18.8	14.6
Plants with 1+ salaried workers			
1988	69.8	17.9	12.3
1990	73.0	15.4	11.6

Source: Central Bureau of Statistics, *Industry and Crafts Surveys*.

Notes: [a] The General Federation of Labour and its subsidiaries; cooperatives; the rural-cooperative sector.
[b] Government; local authorities; the Jewish Agency.

the mid-1980s. This growth, accompanied by the rapid expansion of the government-owned military industrial enterprises, led to the reduction in the proportion of privately owned enterprises in Israeli industry. However, during the late 1980s, a severe crisis emerged in Koor. The concern nearly went bankrupt, and extensive lay-offs and plant closures were required. The number of employees in Koor dropped to around 20,000 and continues to contract as more subsidiaries are sold. Only an extensive bail-out plan approved in 1991 by its creditors seems to have assured the continued existence of Koor in a much reduced form.

The Koor crisis marked the climax of a wider crisis in the Federation of Labour-controlled economy. Management and decision-making that were not based on economic criteria alone led to a situation in which a considerable number of plants were losing money for a prolonged period of time. These plants, particularly those located in development towns, did not close down or restructure, but were subsidised by a small number of profitable (generally high-technology) plants. However, when a severe crisis occurred in the leading electronics subsidiaries, there was no activity left to keep the huge and inflexible concern afloat – particularly as the political environment no longer provided the continued support endowed by the Labour movement. Hence, industrial investment by Federation of Labour enterprises, which were publicly committed to investing in the periphery, reached extremely low levels at the onset of the 1990s. The renewed increase in the proportion of privately owned enterprises in Israeli industry is associated with the crisis in Koor and in the government-owned military industrial enterprises (see Table 12.3).

The rural-cooperative sector, another generator of stable jobs in the periphery, also suffered setbacks during the 1980s. The crisis of the regional agriculture-processing plants, as well as other kibbutz industries, was part of a wider crisis in the agricultural sector and in the kibbutz movement. Difficulties were amplified due to the need to adapt hierarchical patterns of industrial management in a cooperative society committed to values of equality (see Chapter 17). As the kibbutz goes through an ideological change and an economic restructuring process, industry is expected to remain a major economic activity. However, future directions of industrial development in the kibbutz sector are not likely to contribute significantly to the supply of jobs in adjacent development towns.

A quantitative assessment of the repercussions of these shifts on industrial employment in the periphery is yet to be made. The contraction and restructuring of these two sectors present grim prospects for employment growth in the development towns. Nevertheless, it can be strongly argued that these crises merely reflect distortions that were carried on for years, as the necessity to restructure industry in the periphery was delayed in order to prevent unemployment. Hence these crises can also be viewed as a first phase in a process which may lead to the development of a healthier economy in the periphery. A single attempt to reorganise one of Koor's plants in Beer Sheva as a worker-cooperative was successful for some time, but did not provide an example which could be adopted widely (Kislev 1989). Large investments by strong, privately owned enterprises also seem unlikely. Clal Industries, Israel's largest private industrial corporation, also encountered difficulties at the end of the 1980s, although not as severe as those of Koor, due to its more cautious management practices. Foreign investment has also been meagre, although it is perceived as a long-term engine of growth. Hence, small- and medium-sized enterprises may be the most probable short-term candidates for the industrial development of the peripheral areas.

13

THE EMERGING GEOGRAPHY OF HIGH-TECHNOLOGY INDUSTRIES

THE GROWTH OF ISRAEL'S HIGH-TECHNOLOGY INDUSTRIES: AN OVERVIEW

Since the late 1960s, fast-growing high-technology industries in electronics, aeronautics, precision instruments, optical equipment, biotechnology, and subsectors of the chemicals, metal products, and machinery industries, have been perceived as Israel's greatest hope for economic advancement. Israel has lacked significant advantages based on either natural resources (except for potash), access to capital, or geopolitical position. Its only major advantage has consisted of a relative abundance of human capital, making Israel a favourable location for research and development activities and for high-technology industries.

Israel presents a unique case for studying the spatial behaviour of high-technology industries, particularly in electronics. The exceptionally large local demand for sophisticated defence products, combined with abundant human capital, has given Israeli high-technology industries a special place in the international division of labour. In addition to local emphasis on defence industries, a high-technology R&D complex has evolved, in which R&D units of American multinational corporations have played a substantial role. These units have been established in Israel despite the small local market for civilian, high-technology products and the large distance between top management and production units. Led by high-technology industries, the number of engineers and scientists in Israeli industry grew by 460 per cent between 1968 and 1984, and medium-level technicians rose by 300 per cent. At the same time, the overall number of employees in Israeli industry grew by only 50 per cent (Berman and Halperin 1990).

Within Israel, the growth of high-technology industries has had a profound impact on the spatial distribution of industry, leading to a decline in the effectiveness of the government's incentives for industrial dispersal and to a re-emergence of strong pressures to concentrate. More recently, a selective process of dispersal of sophisticated production has begun, influencing the growth prospects of various towns and settlements in the periphery. Attraction of high-technology industries has become a major element in development strategies of local authorities, but only a few localities have achieved success. In this chapter, major phases in the evolution of high-technology industries in Israel are described, followed by an assessment of patterns of spatial concentration and trends of dispersal.

The origins

The origins of Israel's advantage in high-technology industries can be traced to three major factors: a well-developed higher education system; a traditional emphasis on agricultural development; and the extent of national defence needs (Felsenstein 1986). The existence of high-quality educational institutions which benefit from the traditional Jewish emphasis on higher education was central to the creation of Israel's advantage in science-based industries. From its first graduating class in 1931, the history of Israel's institute of technology – the Technion – parallels the history of Israel's technological development in particular (Felsenstein 1986). The great expansion in the supply of qualified technical and scientific labour in Israel's industry was based on the rapid growth of the higher education system during the 1960s and early 1970s, and to a lesser extent on the occupational mix of immigrants arriving after 1967. Direct links between academic research centres and private firms did not develop to a significant extent, and Israel did not enjoy a prominent advantage in the supply of medium-level technical and skilled workers. However, institutes of higher learning, and particularly the Technion and the Weizmann Institute of Science, had a major role in training highly skilled, professional labour, and also served as nuclei for early high-technology agglomerations.

A second factor facilitating Israel's role as a high-technology R&D centre was its long-standing advantage in applied agricultural research and agro-industry (Felsenstein 1986). The emphasis on agricultural R&D has been an ideologically motivated priority of the Zionist movement since the early twentieth century. Israel achieved a leading global reputation in agricultural R&D and became engaged during the 1960s in agricultural development projects in Third World countries. As opportunities for further agricultural development in Israel narrowed, Israelis became increasingly engaged with exporting technological know-how. These exports had a controversial impact on the competitiveness of Israeli agriculture in world markets.

A third – and dominant – element in the development of Israel's high-technology industry was the expansion of its indigenous defence capacity. This expansion was fostered by Israel's large internal market for military products. Of crucial importance for the great push forward in electronics and related industries was the post-Six Day War defence policy that was created in an attempt to expand local capacity in military production. Defence contracts have been said to account for 90 per cent of all local market sales in the electronics industry in the late 1970s, and over 50 per cent of the output in electronics and aeronautics still consisted of military products during the early 1980s (Felsenstein and Shachar 1988).

The growth of Military Industries

The seeds for Israel's military industries were sown in pre-state years (Felsenstein 1986). Military Industries was founded in 1933 as an underground establishment, and evolved during the 1930s and 1940s to operate over forty clandestine armaments plants producing weapons in 'disguised' factories and orchards. After Israel's independence, Israel Military Industries became a subunit of the Ministry of Defence and was expanded and reorganised in a divisional structure, which included a central laboratory and several divisions producing weapons and ammunition. Israel Military Industries began to export some of its products during the late 1950s, including the well-known Uzzi submachine-gun. Employment in Israel Military Industries grew from 2,300 workers in 1950 to 5,916 workers in 1967 (Evron 1980).

Rafael, the Armaments Development Authority, was a second subunit of the Ministry of Defence forming the basis for advanced military production in Israel. Established as an army armaments research unit in 1948, in 1952 it became the Division for Research and Planning (Emet) and was further transformed into its present framework of Rafael in 1958. In 1956, Emet developed the first Israeli-made computer, and later, Rafael was engaged in the development of products such as solid and liquid fuel for rocket engines and anti-tank weapons. A third government-owned defence industrial enterprise was Israel Aircraft Industries, which was established in 1953 by the Ministry of Defence as the Bedek Aircraft Plant. Israel Aircraft Industries began by maintaining, repairing, and modifying aircraft, but gradually expanded its operations until, in 1968, its status was changed from a subunit of the Ministry of Defence to a government-owned corporation.

The French declaration of a Mid-East arms embargo on the eve of the 1967 Six Day War came as a shock to Israel, emphasising the inadvisability of dependence on external arms sources. France was Israel's major arms supplier during the period of 1955–67, and Israel's air force was particularly dependent on French equipment. The embargo played a major role in the decision to give the defence industry top priority to attempt to reduce dependence on foreign suppliers and, utilising its recent intensive warfare experience, to develop technologically sophisticated products which Israel was unable to import. Consequently, the local military industry grew from a base of low, and mostly imported, technology, into a technologically sophisticated military industrial complex, largely involved in electronics, communications equipment, and aeronautics (Felsenstein and Shachar 1988).

Israel's defence expenditures increased from 10.4 per cent of GNP in 1966 to 25.2 per cent in 1980, and the share of the defence sector in the total labour force increased in similar proportions (Mintz 1983). At its height, around 1984, the Israeli defence-industrial sector employed 65,000 workers, including 11,000–12,000 engineers and scientists and 14,000 technicians. R&D expenditures were as high as 33 per cent of the total sales of the defence industrial sector (Halperin 1987), and were estimated at 45 per cent of total national R&D expenditures in 1983 (Hadar 1984). The dominance of the defence sector explained the high rates of R&D investment in Israel, whereas the amount of civilian R&D reached only medium levels compared to other countries.

The military industrial complex included the government-owned enterprises – Israel Aircraft Industries, Israel Military Industries, and Rafael; industrial R&D and production at the Israel Defence Forces itself; other national research institutions; and non-governmental firms. Israel Aircraft Industries became Israel's largest manufacturing enterprise and grew to employ more than 20,000 employees in 1980/1 and 22,000 employees in 1985. Among its wide range of products were the Kfir fighter plane, missiles, and missile boats. Israel Military Industries employed about 14,500 employees during the early 1980s, and Rafael reached more than 5,000 employees. Renovation and development of weapons systems, as well as the assembly line of the Merkava tank project, were conducted within the Israel Defence Forces itself. During the early 1980s, the Merkava project employed 4,000 workers, including a vast network of subcontractors (Mintz 1983).

The first military industrial enterprise not controlled directly by the Ministry of Defence was the Soltam Corporation. It was founded in 1950 as a partnership of Federation of Labour-owned Solel Boneh-Koor and a Finnish corporation and produced mainly heavy mortars. The largest non-governmental defence enterprise, Tadiran, was founded in 1961 through the merger of Tadir (crystal products), owned by Koor, and Ran (batteries), owned by the Ministry of Defence. In the years 1969–72, the Ministry of Defence sold its shares to the American corporation GT&E, with the aim of providing Tadiran with technological and managerial expertise, and particularly with new marketing opportunities. Tadiran, which began by specialising in communications equipment, gradually expanded into a broad array of electronic systems, including equipment exported for electronic warfare and production of electric household goods. At its peak of expansion in 1987, it employed over 12,000 workers. Another joint venture of the Ministry of Defence, this time with the private company Elron, was Elbit, established in 1966. Elbit specialised in developing and producing computer systems, mainly for military applications. As in the case of Tadiran, the government sold its shares in Elbit to an American corporation, CDC, in 1970.

More firms of the Federation of Labour and the private sectors joined the military industrial complex, particularly during the 1970s. Most of the military foreign aid of the US government that Israel received annually was designated for purchases from American corporations; nevertheless, these corporations were permitted to subcontract up to 45 per cent of the total production cost to Israeli suppliers. Hence, Israeli defence industries became heavily engaged as subcontractors to American corporations. The Israeli military industrial complex grew to involve dozens of large companies and hundreds of small subcontractors.

Military production tended to be organised around cost-plus contracts. These lucrative, captive military contracts have been of critical significance to the initial financing of some new firms, reducing their need for venture capital (Friedland 1984). The defence market also had an oligopolistic structure based on a monopsony on the demand side, with the Ministry of Defence being the major buyer, and firms competing on the basis of reliability and performance rather than price (Felsenstein 1986). Since the 1970s, military-based exports have grown substantially and even enterprises such as Rafael have begun to export their products. In 1984, exports reached 38 per cent of the sales of the defence industrial sector, and were some 20 per cent of the national industrial exports (excluding diamonds).

The growth of civilian high-technology industry

The development of civilian high-technology industry has been spawned by defence production. Civilian production in the electronics industry was limited during the 1960s to consumer import-substituting products such as assembling radio sets. Export-based civilian high-technology industry, however, began to grow rapidly during the late 1970s. From the production of advanced military electronics, companies such as Tadiran branched out into exports, first of declassified military items and then of civilian products based on military applications. Among the civilian-oriented enterprises in electronics, most prominent were Scitex (founded in 1968), which began with computerised systems for print design for the textile industry and later specialised in systems for the printing and publishing industry, and Elscint (founded in 1969), which specialised in computer-based systems for medical diagnostic imaging. Both were local entrepreneurial ventures. Elron, the largest high-technology holding company, was developed during the 1960s by Technion graduates with defence R&D experience. Headquartered in Haifa, its major subsidiaries, apart from Elscint, were Elbit, Fibronics (fibreoptics and other high speed information transfer and distribution systems), and Optrotech (computerised electro-optical systems for automation in the printed circuit board industry). Discount Investment Company – a large holding company affiliated with one of Israel's major banks – was also active in joint ventures with some of Israel's early high-technology entrepreneurs, including Elscint and Scitex.

Government involvement in the advancement of civilian science-based industries developed out of the report of the Committee of Inquiry into the Organization and Administration of Government

Research, which, in 1968, recommended the establishment of the Office of the Chief Scientist in the Ministry of Trade and Industry. The role of government in civilian high-technology has increased greatly since 1976/7, largely through grants from the Office of the Chief Scientist to civilian R&D projects with export potential. These grants supplemented the incentives of the Law for Encouraging Capital Investment, for which most high-technology enterprises, as exporters, were eligible. In addition, tax-shelter legislation for high-technology projects (such as the Elscint Law) was in effect for a while during the 1980s.

The growth of civilian high-technology industry has faced several constraints. On the supply side, there has been severe competition from the military industrial complex for scarce qualified labour, which persisted until the slow-down in the defence industries in the mid-1980s. In addition, it has been argued that high-technology start-ups have a difficult time raising capital in Israel due to lack of venture capital (Felsenstein and Shachar 1988). Government support for high-technology investment has tended to find its way to the largest firms and holding companies, and small enterprises have been recipients of only about 10 per cent of all Israeli government aid. During a period in the 1980s, the Elscint Law allowed tax deductions for personal investments in high-technology securities. However, this law was restricted to firms which had already exported $20 million worth of goods based on their own R&D (Friedland 1984), thus ruling out any support for new start-ups.

On the demand side, local civilian demand for high-technology products was very limited. Hence, growth of civilian high-technology industry could not be based on easily accessible local markets. Whereas foreign sales by American companies followed a company's success in establishing itself in domestic markets, Israeli companies frequently sold their first domestic products only after achieving significant export sales. Companies such as Scitex were 100 per cent exporters for years, their products being too expensive even for the few potential Israeli customers.

The role of multinationals

The early 1970s saw accelerated entry of American high-technology firms into the Israeli electronics industry. Israel was generally unable to put together a more attractive package of incentives than other countries (for example, Singapore or Ireland), nor could it offer particularly cheap labour. Labour expenses per employee in the Israeli electronics industry were about one-half those in the same industry in the United States in 1985, but were similar to those in Japan and Western European countries and higher than those in Southern Europe and South-East Asia (Ministry of Industry and Trade 1990). However, the ability to put together highly skilled R&D teams that foreign companies were unable to assemble at home was considered a major Israeli advantage for multinational corporations.

Israel offered stable, top-quality scientific labour, which was less likely to 'job-hop', and at a slightly cheaper cost than in the United States (Felsenstein and Shachar 1988). In addition, Israel could offer experience gained through military R&D. Government incentives also had a role in offsetting the political risks of investing in Israel and in gaining the competitive edge over other countries. A location in Israel enabled firms to take advantage of reciprocal trading ('buy back') agreements between Israel and the US, by which Israeli purchases from US companies had to be compensated for by purchases in Israel by US firms. These purchases were then covered by offset credits to the firm by the US government (Christopherson and Gradus 1987). Israel's agreements with the EEC were also an advantage in attracting American multinationals. Thus, Israel developed, to some extent, a unique role in the international division of labour, attracting R&D and skill-intensive production units of multinational firms in the electronics industry headquartered in the United States (Christopherson and Gradus 1987).

The role of cultural differences in creating Israel's advantage has been emphasised by Dov Frohman, the general manager of Intel Israel. According to Frohman (1989), the American frontier culture was based on the 'gold rush' legacy, and has focused on individualism, privacy, personal responsibility, and hierarchical management. By contrast, the Israeli frontier culture has been linked to external security threats, and has focused on collective action and responsibility, openness, informality, and less respect for authority. These differences have been associated with differences in spatial and job mobility, which are both much lower in Israel than in America. Together with the deeper attachment to family and local community, and with the relative lack of venture capital, Israeli teams of professional employees have been assumed to be more stable than their American counterparts. While Israeli culture has been more and more influenced by that of the US, particularly amongst that professional stratum engaged in high-technology development work, the cultural differences still offer unique opportunities for American multinational corporations.

The penetration of multinational corporations into the Israeli electronics industry stood in contrast to the small share of foreign companies in the Israeli economy. Foreign-owned companies were estimated to account for only 1 per cent of total employment in Israel, in comparison with 33 per cent in Belgium and 19 per cent in Scotland (Jerusalem Institute of Management 1987). The Arab boycott limited foreign investment in Israel, adversely influencing mass-production of final products in particular. Most foreign investment was traditionally made by Jewish investors, frequently not in the framework of large multinational corporations. The case of high-technology industries, particularly electronics, therefore represents a break with earlier trends.

Around 150 US corporations have established subsidiaries involved in some form of R&D in Israel, or have subcontracted for R&D with an Israeli firm or another US–Israeli subsidiary (Christopherson and Gradus 1987). The form of investments that foreign companies have made can be broken down into four categories (Felsenstein 1986). First, foreign companies have established subsidiaries to develop a prototype in Israel, with the production and marketing being done in other countries. Several major US semiconductor companies have established R&D and design centres in Israel (Motorola in 1964, Digital in 1973, Intel in 1974, National Semiconductor in 1978, Daisy Systems in 1984). Second, foreign companies have established subsidiaries to develop a prototype and manufacture it in Israel. A successful operation of an R&D unit has frequently led to local expansion through the establishment of production units. Such was the case with Intel and National Semiconductor. In some cases (for example, Motorola Communications), the production unit has close links with the R&D unit. In other cases (for example, Intel) there have been no significant direct links, and the existence of the R&D unit mainly served to highlight the Israeli option to the mother corporation and to reduce uncertainties associated with it. Third, joint ventures with local companies can be set up. During the early 1970s, GT&E and CDC became shareholders in Tadiran and Elbit, respectively (both are no longer involved in these companies). Finally, a less involved relationship might be investment in R&D projects in Israel or agreements to market Israeli high-technology products abroad.

Thus, unlike typical patterns of expansion for American high-technology multinationals in South-East Asian countries, such corporations did not penetrate Israel by establishing production units and later expanding into more sophisticated activities; rather, a common entry into Israel was through R&D or marketing units (IBM established a marketing, services, and maintenance subsidiary in Israel as early as 1950). The successful operation of an R&D unit could contribute to a decision to expand local operations to include relatively advanced phases of production. American multinationals also bought shares in existing Israeli defence-related industrial enterprises in order to benefit from their experience and to gain access to the local market, while simultaneously opening new marketing and financing opportunities for the Israeli partner. Israeli subsidiaries of American corporations offered many well-paid professional job opportunities: in 1987, for example, more than 50 per cent of Motorola's employees (1,903 in 1987) were engineers, academics, and technicians and over 30 per cent of the sales of the Israeli operations of Motorola were comprised of products based on local R&D. Of Intel Israel's 870 employees in 1989, more than 50 per cent were engineers, academics, and technicians.

Reservations as to the wisdom of attracting large US corporations to Israel and of the effect such corporations might have on Israel's future innovative capacity were raised by Friedland (1984). The large multinationals play the competitive game between countries for aid and subsidies; Intel, for example, negotiated with Ireland and Singapore until it finally decided to build in Jerusalem. Israeli firms can also sell out to major US firms to get access to the world market and to get the capital denied them within Israel. As ideas developed in Israeli subsidiaries are converted to assembly line production, international competition can force the plants of US multinationals to locate outside of Israel. Israeli firms, on the other hand, are more likely to scale-up production within Israel. Intra-corporation competition with South-East Asian units and geopolitical instability in the Middle East pose some threat to Israeli units of multinational corporations. Multinationals could also drain the local manpower and thereby inhibit local small-scale innovative activity (Friedland 1984).

Yet, during the world-wide slump in the electronics market in the mid-1980s, branch plants of multinationals proved to be more stable than plants of Israeli companies. Their advantage over locally owned enterprises lay in not being restricted to specialised niches in which international competition of giant corporations was limited.

Major integrated circuits producers in Israel have been American multinationals, while Israeli companies have capitalised either on their advantages in defence-related products, or have specialised in niche markets neglected by the large multinationals.

Israeli companies have also faced constraints in marketing. The severe difficulties encountered by Tadiran, after the American corporation GT&E sold its shares to its Israeli partner, Koor, demonstrated the difficulties of Israeli corporations which lack the backing of foreign multinationals in marketing efforts. The ambitious purchase of a marketing system in the United States led to the crisis in Elscint. It would seem that the possible reorganisation of international trade into regional blocks, associated with the economic unification of Europe in 1992, can only reinforce the advantages of integrating local high-technology operations with multinational enterprises.

Crisis and restructuring

The period of rapid growth in Israel's leading high-technology industries terminated, at least temporarily, in 1985, revealing the perils of overreliance on this sector, which faces dynamic and highly competitive world markets and political constraints in exporting military equipment. The crisis was primarily a product of the convergence of two separate processes: a world-wide downturn in the electronics market and cuts in Israel's defence budget. These cuts became most significant with the government's successful efforts in 1985 to stabilise the economy and put an end to the three-digit inflation levels. The crisis perhaps reached its climax with the government's retreat from the Lavi fighter-plane project in 1987. The R&D phases of the Lavi project engaged 4,500 workers, who accounted for more than 10 per cent of the supply of technological labour in Israel (Berman and Halperin 1990). The retreat from the Lavi had a negative impact on the transport equipment industries (see Figure 2.4). Israel Aircraft Industries had to cut its workforce by one-quarter, from over 22,000 in 1987 to 16,617 in 1989, in a successful restructuring effort. Moreover, numerous subcontractors in other enterprises were also affected.

It became more evident that the over-dependence of Israeli high-technology industries on military demand negatively influenced prospects for expansion and consolidation of the unique position held by Israel in the international division of labour. Defence industries faced two major limitations to growth (Berman and Halperin 1990). First, the local defence demand facilitated the growth of sophisticated industry, but only to a certain point, beyond which its expansion prospects were restricted. The industry grew too large for local demand, and dependence on local markets made it extremely vulnerable to defence budget cuts, leading to instability. Israeli defence industries concentrated on creativity, modification, and filling niches within the markets, thereby avoiding costly stages of design, copying, rebuilding, etc. (Felsenstein 1986). However, as the defence industries reached a certain size and level of sophistication, further growth would have meant dealing with larger and more ambitious projects, such as jet-fighters, in which development costs would have been extremely high. While Israel had the technological capability, it did not possess the economic strength to finance such projects, and the size of its local defence demand did not provide sufficient economies of scale. This was reflected by ever-increasing R&D expenditures in the defence industry, up to the point where local sales and government support could not sustain them anymore.

A second constraint concerned export markets. Export of defence products began to grow rapidly after 1973, and export policies with respect to defence systems gradually liberalised. However, the growth of defence-related exports slowed down during the 1980s. Exporters of defence products faced economic and political instability in countries such as South Africa, Iran and South America. Exporting weapons involved particular political risks for a small country such as Israel, which did not even have diplomatic relations with many countries of the Third World and the Eastern Bloc. In some cases, Israel's geopolitical position pushed local companies to export to dubious clients and regimes. In addition, constraints existed with respect to exporting more sophisticated systems, which was considered a security risk. On a broader scale, it has also been argued that countries with large military budgets have exhibited slower rates of economic growth. Military contractors absorb available capital and the best brainpower, diverting them from civilian industrial R&D which could potentially lead to far greater growth (Friedland 1984).

Civilian high-technology industries suffered from the lack of local markets and distance to major foreign markets, augmented by Israel's perceived geopolitical position. Lack of venture capital has also been quoted frequently as an obstacle to growth of civilian high-technology enterprises in Israel. Expectations raised during the early 1980s that Israel's software industry would reach $1 billion worth of exports by the end of the decade did not materialise, and the software industry remained dependent on local markets.

The global downturn in the electronics industry during the mid-1980s had a particularly strong influence on Israeli enterprises. Israeli high-technology companies are usually small relative to their competitors in the major industrialised countries. As noted above, they could generally compete successfully in small specialised niches which did not attract the larger corporations. However, if such niches became attractive to larger corporations, and the Israeli companies were not flexible enough to adjust or to retain a technological edge, the Israeli companies would run into difficulties. Two of the 'flagships' of Israeli civilian high–technology industries – Elscint and Scitex – underwent severe crises. Elscint, whose problems began in 1984, was rescued by massive government support. Scitex encountered difficulties one year later and was more successful in recovering, regaining profitability and retaining its prominent position in world markets.

More specific factors contributed to the crisis in high-technology industries, such as the devaluation of the US Dollar relative to European currencies, the growing difficulties in financing R&D and working capital, and factors endogenous to specific firms (Teubal 1989). The crisis has been clearly reflected by data for the electrical and electronic equipment and the transport equipment branches (see Figure 2.4). Separating defence high-technology industries from civilian high-technology reveals that the core of the crisis was in the defence sector (Berman and Halperin 1990). While defence industries contracted in absolute terms, the growth of civilian high-technology enterprises continued, although at a slower pace than prior to 1985.

Utilisation of the advantage of Israel (and world Jewry) in medical sciences to develop a nucleus of a biotechnology industry has proved so far to be more difficult than creating an electronics complex, and local attempts to gain a foothold in the expending biotechnology industry have met with modest success. Biotechnology did not enjoy the strong backing of local military demand. In addition, unlike electronics, global R&D in biotechnology is dominated by large multinationals, and barriers-to-entry have been high for many products, largely due to the extremely long time gap between the initial development and final approval by national

THE EMERGING GEOGRAPHY OF HIGH-TECHNOLOGY INDUSTRIES

health authorities. These conditions make scientist-entrepreneurs in biotechnology dependent on large corporations and venture capital for long-term financing and marketing. Multinationals in these fields are reluctant to conduct business in Israel due to the Arab boycott. Nevertheless, Israel is able to offer expertise in this field, particularly at the Weizmann Institute in Rehovot and the Hebrew University of Jerusalem, and through its sophisticated pharmaceutical industry. Bio-Technology General, for example, has been primarily engaged in genetic engineering and related R&D since it commenced operations in the early 1980s at a location near the Weizmann Institute, and it had 130 employees in 1989. Nearly thirty biotechnology enterprises, employing over 1,000 researchers, operated in Israel in 1991.

THE EVOLUTION OF MAJOR HIGH-TECHNOLOGY SPATIAL CLUSTERS

Major trends in the spatial pattern of high-technology industries in Israel have consisted of an initial concentration in Tel Aviv and its vicinity, followed by the evolution of secondary cores around institutes of higher learning, particularly in Haifa and Rehovot. At a later stage, a selective process of dispersal began. Nevertheless, the Tel Aviv metropolis has retained its dominance, and has not yielded its position to newer planned high-technology concentrations or to dispersal processes.

The pre-1967 seeds of the electronics industry, which consisted of military-related enterprises and of plants engaged in import-substituting consumer electronics, tended to concentrate in the Tel Aviv metropolitan area (see Table 6.1). Links with universities were perceived as a major location factor during the post-1967 years of growth. Several science-based firms were established during the mid-1960s in the Technion in Haifa and in the Weizmann Institute in Rehovot. The Ministry of Commerce and Industry decided, as early as 1968, to support the establishment of four centrally located, high-technology industrial parks near Israel's major institutions of higher learning in Haifa, Rehovot, Tel Aviv, and Jerusalem, financing nearly the total investment necessary for their development (State Comptroller of Israel 1982).

The spatial distribution of employment in high-technology industries in 1969/70 presented in Figure 13.1 is based on thirty-nine firms, employing a total number of 6,366 workers. Tel Aviv and its environs (Rehovot, Lod) accounted for nearly 50 per cent of all

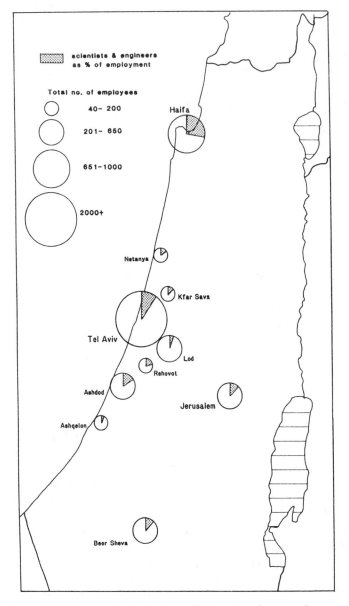

Figure 13.1 The spatial distribution of high-technology employment, 1969/70

Source: Felsenstein (1986: 46).

high-technology employment. Haifa was the only noticeable secondary area. The first industrial science park in Israel was established there in 1969, and a major complex of new high-technology enterprises emerged during the 1970s. Proximity to the Technion was a critical factor in the early development of the Haifa electronics complex, while proximity to the Weizmann Institute was responsible for the establishment of the nearby Weizmann Science Park. The Hebrew University in Jerusalem became a less prominent nucleus of high-technology development, emphasising mostly pharmaceuticals; nevertheless, the Tel Aviv metropolis remained the most dynamic core of high-technology activity (Felsenstein 1986; see Table 6.1).

The spatial distribution of high-technology industries in 1982/83 (see Figure 13.2) revealed similar patterns to those of 1969/70 (see Figure 13.1), despite the rapid growth of high-technology in the interim period (Figure 13.2 is already based on data for 346 plants and 51,696 workers). In the light of immense growth, some dispersal was inevitable, but the Tel Aviv, Haifa, and Jerusalem metropolitan areas still accounted for 73.2 per cent of all high-technology plants and 78 per cent of all employment, with the Tel Aviv metropolitan area alone responsible for 52 per cent of all jobs (Felsenstein 1986).

High-technology industries in the Tel Aviv metropolis evolved mainly in an unplanned manner. Large enterprises which tended to locate in the city of Tel Aviv until the 1960s showed an increasing preference for suburban locations, whereas small-scale operations remained concentrated in the inner parts of the metropolis. Hence, software companies, data processing, and computer-related services have been mostly concentrated in the city of Tel Aviv and in the inner suburb of Ramat Gan (see Table 13.1). Within the suburbs some unplanned high-technology clusters have emerged (see Figure 13.3). Most prominent in its diversity, if not in number of employees, has been the Herzliyya industrial estate, which became a major nucleus of private high-technology firms, including all operations of Scitex and R&D and service centres of American corporations (Digital, National Semiconductor, and Daisy Systems). Petah Tiqwa has been another sub-centre, where the Federation of Labour-owned high-technology enterprises, led by Tadiran Communications, have been concentrated. The core of Tadiran's activity has been in the large and diversified Holon industrial estate (Grossman *et al.* 1983). The Ben-Gurion Airport–Yehud–Lod complex has

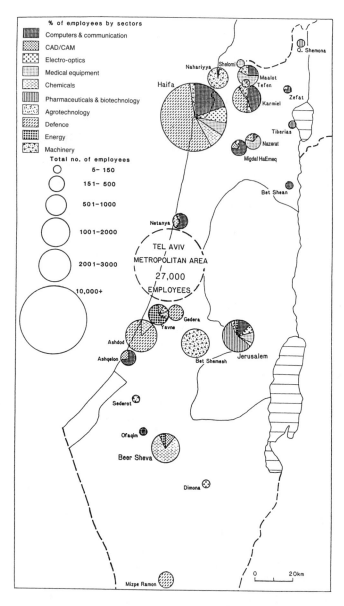

Figure 13.2 The spatial distribution of high-technology employment by industry, 1982/3 (not including small plants in rural settlements)

Source: Felsenstein (1986: 56–8).

Table 13.1 Firms in selected high-technology manufacturing industries by location, 1990[a]

Location	Computers and computer peripheral equipment	Aerospace, shipbuilding, missiles, tanks and transp. equip.	Precision instruments, optical equip. etc.	Medical, electromedical and related instruments	Communications equipment	Semi-conductors and electronic components	Batteries and miscellaneous electrical equipment	Software, data processing and computer-related services
Tel Aviv metro. – total	69.8	36.4	50.4	56.1	70.1	59.1	61.8	82.5
Tel Aviv	25.6	7.3	17.9	20.7	25.3	16.5	18.4	39.9
Inner suburbs – East	18.6	1.8	4.9	8.5	16.1	4.7	17.1	24.2
Outer suburbs – East	7.0	14.5	4.1	7.3	13.8	12.6	10.5	5.7
Suburbs – North	11.6	7.3	7.3	8.5	3.4	10.2	3.9	9.7
Inner suburbs – South	4.7	5.5	8.1	7.3	11.5	10.2	10.5	1.7
Outer suburbs – South	2.3	0.0	8.1	3.7	0.0	4.7	1.3	1.3
Haifa Metropolitan Area	9.3	12.7	13.0	9.8	6.9	11.8	14.5	7.4
Jerusalem and Mevasseret Ziyyon	0.0	1.8	8.1	7.3	2.3	5.5	3.9	4.0
Central Coastal Plain	0.0	7.3	4.1	3.7	3.4	2.4	1.3	0.7
Western Galilee	9.3	10.9	7.3	2.4	1.1	5.5	3.9	0.7
Eastern Galilee and Golan Heights	0.0	3.6	0.8	0.0	1.1	3.1	2.6	0.0
Jewish settlements in the West Bank	0.0	1.8	0.0	1.2	1.1	1.6	0.0	0.0
Southern Coastal Plain	2.3	9.1	2.4	4.9	4.6	2.4	2.6	0.7
The South	4.7	1.8	0.8	0.0	2.3	0.8	2.6	1.3
Rural Settlements[b]								
Eastern Galilee and Golan Heights	2.3	14.5	13.0	14.6	6.9	7.9	6.6	2.3
Western Galilee and Haifa Area	0.0	7.3	6.5	6.1	4.6	1.6	1.3	0.3
Central Coastal Plain	0.0	3.6	3.3	4.9	2.3	3.1	1.3	1.0
Southern Coastal Plain	0.0	3.6	0.8	1.2	0.0	2.4	0.0	0.3
The South	0.0	0.0	0.8	0.0	0.0	0.0	0.0	0.7
	2.3	0.0	1.6	2.4	0.0	0.8	3.9	0.0
Total (absolute numbers)	43	55	123	82	87	127	76	298

Source: Compiled from data published in *DunsGuide 1991*, Tel Aviv: Dun & Bradstreet (Israel) Ltd.

Notes: [a] The Table includes only firms listed in *DunsGuide* and does not include branch plants which are not registered as separate companies. Firms operating in more than one industry may appear in more than one column.
[b] Kibbutz, Moshav and small Yishuv Kehillati.

been the core of Israel Aircraft Industries activities, in addition to the Federation of Labour-owned Telrad (see Figure 13.3).

Two of the four original planned science parks, initiated with government support near Israel's institutions of higher learning, have been in the Tel Aviv metropolis. One is the Atidim Science Park (in the northern part of the city of Tel Aviv). The second, which is the major planned science park in the Tel Aviv metropolis, is the Rehovot-Nes Ziona Weizmann Science Park.

The initiative for the establishment of the Qiryat Weizmann Science Park came from the Weizmann Institute during the late 1960s. Pinhas Sapir linked the Weizmann Institute with the Africa–Israel building and real-estate company, which became developers and managers of the park. Sapir also helped to persuade El-Op to move from Tel Aviv to the new park as a large leading enterprise. The first building in the park was completed in 1971, and the first four tenants were small firms that had emerged within the Weizmann Institute. During the early 1970s, it was difficult to find enterprises expressing interest in the park, and the building stood empty for some periods. Only towards the late 1970s did the science park begin to develop rapidly (Felsenstein 1985). In 1985, about 58 per cent of the 2,400 workers in the Weizmann Park were in electro-optics and 15 per cent were in biotechnology.

Major location factors of Qiryat Weizmann enterprises were place of residence of owner and proximity to the Weizmann Institute. However, only a few enterprises (in biotechnology and life sciences) kept significant linkages with the Weizmann Institute. For most firms, the proximity to Weizmann was only important for the image. Tenants in the Weizmann Science Park complained of high rents, and a move to localities such as Yavne or Qiryat Malakhi was considered an option to reduce costs while enjoying a fairly similar location (Felsenstein 1985).

Haifa formed the second major high-technology cluster in Israel (see Figure 13.2 and Table 13.1). Its core, the MTM Science Industries Center, has been the largest single site of private high-technology industries in Israel and has included tenants such as Elbit, Elscint, Fibronics, and Intel. Another dominant enterprise in Haifa has been the government-operated Rafael, which played, on a smaller scale, a role similar to that of Israel Aircraft Industries in the Tel Aviv metropolis. The roots of Haifa's high-technology complex were closely linked to the Technion, and some of the early ventures were associated with personalities from that institution. The

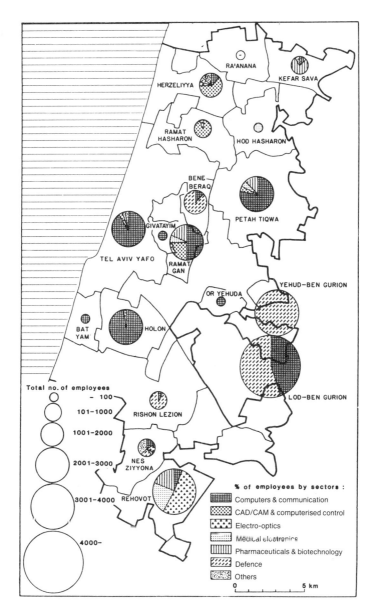

Figure 13.3 The spatial distribution of high-technology employment in the Tel Aviv metropolitan area by industry, 1982/3

Source: Felsenstein (1986: 55).

Technion had an additional and perhaps greater role: it is widely perceived as a major location factor, and thus serves to attract public support for Haifa's science park. Residence in the Haifa area and education at the Technion were quoted as major location factors by small high-technology firms in Haifa, while the availability of premises for immediate occupation and the availability of skilled R&D labour ranked next in importance (Shtainmets 1989). These last two factors have been associated with the agglomeration economies developing in Haifa, thus maintaining and reinforcing Haifa's advantage.

Quality of life was also quoted to be central in the evolution of Haifa's high-technology complex. It was argued that the success of Haifa's science park was related to its proximity to prestigious neighbourhoods on Mount Carmel, with one of Israel's best high schools, and to Haifa's top-level cultural amenities, public services, and attractive physical setting, such as the bay and mountainous terrain (Rapaport 1984). However, the perceived role of quality of life has probably been exaggerated (for similar cases, see Scott and Storper 1987), and it was not ranked as a significant location factor by high-technology firm owners and managers (Shtainmets 1989).

It should be emphasised that while often considered as Israel's premier example of a high-technology complex, Haifa never rose to a prominent position, and it has always been in the shadow of Tel Aviv. In fact, high-technology enterprises in Haifa depended on the Tel Aviv metropolis in some respects. In the software industry in particular, a heavy dependence on Tel Aviv is often expressed via the establishment of marketing offices in the Tel Aviv metropolitan area by Haifa firms. Furthermore, enterprises in Haifa depend on specialised business services from Tel Aviv and even depend to a certain extent on its labour market for top-level academic employees. Only 67 per cent of the top management academic labour in Haifa's high-technology industries resided in the Haifa metropolitan area in 1987, whereas 16.2 per cent commuted from the Tel Aviv region (Shtainmets 1989). Thus, the location of the MTM Science Industries Center near the Haifa–Tel Aviv freeway has been most advantageous, given the strong links of local enterprises with the Tel Aviv area.

High-technology formed only one part of Haifa's industrial image, the other one being that of an old and declining industrial region. The Haifa metropolis has been in a process of retreat in Israel's urban system (Haifa and Galilee Research Institute 1989).

The competitive position of its port has deteriorated in comparison with the more centrally located Ashdod Port, and Haifa has suffered from its marginal position in the national transportation system. Its role as a major industrial centre has gradually waned in favour of the Tel Aviv area on the one hand, and development towns in the western Galilee on the other. Commercial activities have penetrated Haifa's industrial areas, and it can no longer offer significant location advantages, such as availability of land, cheap labour, government subsidies, or central location. Petrochemicals and electronics have remained the major thriving industrial complexes in the Haifa area.

The Federation of Labour-owned enterprises had a major role in the formation of the industrial base in the Haifa area during the 1940s and 1950s. But although the high-technology electronics industry developed rapidly in Koor (the major industrial corporation of the Federation of Labour) during the 1960s and 1970s, this expansion concentrated in the Tel Aviv metropolitan area, mainly in Koor's subsidiary, Tadiran. The lack of R&D-intensive plants owned by the Federation of Labour enterprises in Haifa has been striking, and can be partly attributed to the political struggle leading to the transfer of Koor's headquarters from Haifa to Tel Aviv and to the retreat of the Federation of Labour enterprises from Haifa, discussed in Chapter 6. Thus, a clear distinction has emerged between Haifa's old industrial base, led by Federation of Labour enterprises, and the new privately owned high-technology firms. These new firms were not discouraged by Haifa's strong labour movement, since they were oriented to a completely new segment of the labour force which was non-unionised and showed little solidarity with Israel's established labour movement.

Jerusalem evolved as a third, much smaller, concentration of high-technology industries (Felsenstein 1988a, 1988b). These industries were perceived as suited to Jerusalem across a variety of criteria. They were thought to be space-efficient, environmentally clean, job-creating activities that could capitalise on the local human capital and local research institutions. It was further felt that the lack of industrial heritage in a city dominated by public services would not adversely affect the development of high-technology industries, since these represented a 'clean break' from traditional industry. A scientific and research infrastructure existed in the city with a basis in the health and biological sciences (The Hebrew University) and, to a lesser extent, in electro-optics and computer sciences (Jerusalem College of Technology). Hence, Jerusalem was one of the two main centres of biotechnology firms in Israel. Of a sample of twenty-one biotechnology enterprises in 1991, six were located in Jerusalem, and six in Rehovot, the other major cluster of this industry.

Jerusalem's Har Hotzvim Science Park was a joint initiative of the Hebrew University and the municipality and managed by the government-owned Jerusalem Economic Corporation. The first building in Har Hotzvim was completed in 1971. During the early phases of the development of the park, the association with the university gave it a prestigious image. Later on, however, the university ceased to be a particular asset. The park and its tenants enjoyed preferred development zone incentives, which remained in effect during the 1980s when science parks in the Tel Aviv and Haifa metropolitan areas were no longer entitled to preferred incentives (Felsenstein 1988b). Teva, Israel's largest manufacturer of pharmaceuticals, was the leading enterprise in Har Hotzvim. In 1982/3 it included nearly one-half of Jerusalem's high-technology employment (see Figure 13.2). Additional large plants were established in Har Hotzvim during the 1980s—Luz (solar electric generating systems), which closed down in 1991, and Intel (semiconductor fabrication), which became the largest employer in the park. In 1987, 45 per cent of the 1,800 workers in Har Hotzvim were in electronics and 24 per cent in chemicals (Felsenstein 1988a).

Unlike in the other metropolitan areas, the government has been the principal agent in encouraging the growth process in Jerusalem. Government incentives are quoted by firms as the major location factor, followed by availability of local academic labour and place of residence of the founder. As in the case of Haifa, quality of life was not ranked as an important factor. A high rate of local firms expressed disappointment concerning what they perceived as Jerusalem's advantages. On the one hand, the government functioned as the principal actor in spawning the growth process, but on the other hand, the government was perceived as a major deterrent to the growth process, setting up a formidable bureaucracy that stifled latent entrepreneurial potential. The experience of plants locating in Jerusalem indicated that Jerusalem did offer sufficient labour of academics and engineers, but the supply of intermediate-level skilled labour was more of a problem, since Jerusalem's economy was dominated by public services and lacked an industrial tradition (Felsenstein 1988b). High-technology industries in Jerusalem did not form a complex in the full sense of the word. A large proportion of firms (42 per cent) reported no linkage at all with other companies in Jerusalem, and most had weak or minimal local linkages (Felsenstein 1988b). Links with local academic institutions were also weak, and involved employment of university graduates and use of consultancy services, rather than joint funding, research projects, etc. Like Haifa, Jerusalem's enterprises were also dependent on Tel Aviv for top-level business services.

Other centres of sophisticated production (see Figure 13.2) have usually not been much more than the sum of one or two large plants. Ashdod and Yavne are practically on the fringes of the Tel Aviv metropolis. Beer Sheva has been a noticeable cluster in the periphery, but it consists mainly of the chemicals industry and not of leading electronics. A local initiative in the field of computerised systems, which was established in Beer Sheva in 1967, did not serve as a nucleus for further development, and during the 1980s this firm encountered difficulties, contracted, and moved its operations to central Israel.

ATTEMPTS AT DISPERSAL

For more than a decade, high-technology facilities were able to receive incentives in metropolitan areas equivalent to those of the highest development zone status. Attempts to promote their dispersal faced constraints derived from the strong tendency of knowledge-intensive industries, characterised by complex linkage patterns, to concentrate in a few spatial agglomerations (Scott and Angel 1987). When preference was finally granted to development towns, it did not offset the disadvantages of remote location. Officials warned that increased land costs and reduced capital subsidies in central areas might provoke high-technology firms to transfer their activities overseas rather than to development towns, since the latter lacked adequate professional manpower and amenities (Rapaport 1984). Advanced defence industries had very little tendency to disperse, despite being affiliated with the government (see Chapters 10 and 12).

Since the late 1970s, the maturing of some high-technology enterprises and reduced public support to such plants established in metropolitan locations led to the dispersal of some branch plants of high-technology firms into a ribbon of development towns within commuting distance of Haifa: Maalot, Tefen, Karmiel, Nazerat Illit, and Migdal HaEmeq (see Figure 13.2). This string of development towns east of Haifa has enjoyed favourable location and highest development zone status, and since the mid-1970s has developed faster than any other region of development towns (Razin 1986). Several high-technology industrial plants were also established in Qiryat Malakhi and Sederot, which enjoyed similar advantages of preferred incentives in addition to their location within commuting distance from the southern fringe of the Tel Aviv metropolis. In addition, attempts have been made since 1981 to promote high-technology industries in Ariel – one of the major Jewish settlements in the occupied territories. Ariel enjoys an 'A' development zone status and is in closer proximity to Tel Aviv than any settlement of this classification within Israel's pre-1967 boundaries.

However, high-technology plants established in these development towns have usually been branch plants involved in products and processes of the later stages of the product life cycle. Most firms engaged in R&D have been located in the centre of the country. While nearly 40 per cent of the employees of high-technology enterprises in metropolitan areas were scientists, engineers, and technicians, for non-metropolitan areas this figure was only 24.4 per cent (Felsenstein 1986). The majority of plants located in development towns have been relatively large production facilities, with a lower demand for skilled labour and a tendency towards a capital intensity in relation to labour (Shefer and Frenkel 1986). The non-metropolitan plants expected to benefit from subsidised capital, reduced taxes, and less expensive labour available in the development towns, while skilled labour could still commute from metropolitan areas or from 'high quality of life' rural settlements. A severe shortage in industrial land in the northern coastal plain also motivated the dispersal from the Haifa metropolis eastward (Sofer 1976).

Yet the limited dispersal of high-technology industries, evident in Figure 13.2, has mainly been a function of the immense growth of these industries. This has led to some marginal growth in the periphery, while the concentration of these industries in the central metropolitan areas has remained essentially unchanged. Nevertheless, a new spatial specialisation emerged among the non-metropolitan development towns. In the north, enterprises engaged in significant R&D activities tended to disperse from Haifa to Migdal HaEmeq and Yoqneam – that is, in closest proximity to the metropolitan area. Plants locating further from the metropolis, such as in Karmiel, tended to be branch plants with little R&D activity

(Shtainmets 1989), whereas hardly any high-technology facilities dispersed to more remote towns in the eastern Galilee.

Other evidence of sophisticated production in the periphery has been in kibbutzim. The locational patterns of kibbutz industry have differed from those of other sectors, since location is tied to the 'parent' kibbutz, and the backing of a nation-wide cooperative structure can compensate somewhat for peripheral location. Kibbutz enterprises have mostly been small-scale, single site activities in the fields of electronics, agro-technology and the more simple end of the biotechnology market, such as fermentation processes and the production of plant tissue cultures. The kibbutz, however, has been prominent in the adoption of high-technology production processes such as plastics, pick-and-place machinery, and CNC (computer numerically controlled). One of the main drawbacks of advanced kibbutz industry has been the lack of skilled manpower in sufficient numbers among kibbutz members to allow their plants to innovate and grow.

A skilled workforce, particularly one at the highest academic levels, has constituted a major consideration in locational decision-making for high-technology enterprises in Israel (Bar-El and Felsenstein 1989). The complex input, output, and information linkage patterns of young R&D-intensive enterprises could have also played a most significant, even if less measurable, role in inhibiting their dispersal.

An exception to the typical dispersal processes of the 1980s has been the industrial incubator park facility initiated in the Tefen industrial area, near Maalot in the north of the country, by Stef Wertheimer – a dynamic entrepreneur (Weitz 1989). The incubator facility – called Ganei Taassiya (industrial garden) – has been a carefully planned and managed combination of industrial space, central services, and an open-air museum. Ganei Taassiya opened in 1984 and housed nineteen firms by the end of the 1980s, employing about 250 workers. It has been very active in public relations, promoting the values of innovative entrepreneurship and development of export industries, and criticising the government's intervention in the economy. Wertheimer is the founder of Iscar, an advanced producer of precision cutting tools for metal and wood established north of Haifa in Nahariyya in 1952. The incubator park project is part of a wider Wertheimer-led initiative aimed at promoting industrialisation north of Haifa. It has included the transfer of Iscar Corporation from Nahariyya on the northern coastal plain to the rural Tefen region further inland, and the establishment of an adjoining project – a high-quality-of-life village of single-family houses – called Kefar HaVeradim (Rose Village). However, despite the attractive location, and the 'A' development zone status, this initiative has grown slowly (Prister 1987). The semi-peripheral location has apparently been problematic for those smaller enterprises that depend on complex and irregular input–output and information linkages. In addition, not all of the tenants in the incubator park conform with the formal criteria for acceptance. Incubator facilities for advanced industries have subsequently became a popular economic development slogan for local authorities, but despite plans to reproduce the incubator park concept in other locations, growth has been very slow.

This limited dispersal of high-technology industry in the north of the country was not replicated in the south. The relatively high figure for electrical and electronic equipment industrial activity in the Southern District in 1972 (see Table 6.1) is misleading, since 92 per cent of this figure merely represents suburbanisation of large branch plants to Ashdod in the fringe of the Tel Aviv metropolis. By 1983, of the sector's 2,700 employees in the Southern District, only 200 were in the remote south. A severe economic crisis affected the southern development towns over the 1970s and 1980s. The Negev area was replaced in the national scale of priorities by the West Bank and the Galilee, where the territorial conflict between Jews and Arabs was more evident (Sofer 1986).

Essentially no new plants were established in Ofaqim, Dimona and Yeroham, although these small southern towns were granted highest development zone status. Even the well-established towns of Beer Sheva and Arad stagnated, being unable to compete with other towns offering the same incentives but having a more central location and a more attractive physical setting than that of the Negev desert. The strength of the mineral-extracting plants, particularly of Dead Sea Works, and the stable employment offered by the Center for Nuclear Research near Dimona, together with Ben-Gurion University of the Negev at Beer Sheva, and other regional functions in that city, saved the Negev from even more severe employment contraction. Those sophisticated plants that were attracted to the Negev were mainly polluting industries or those engaged with treatment of toxic materials, for which the Negev was their only possible choice. Being a closed self-sufficient system, the Center for Nuclear Research could not act as a core for high-technology development. Small advanced firms in Beer Sheva faced severe constraints due to the lack of qualified local labour. This difficulty stressed the missed opportunity of not locating a significant portion of the military industries in the Negev. Unlike small high-technology ventures, such large organisations have not been limited to local labour markets, and could have recruited labour nationally, thus changing the characteristics of the local labour market.

The location behaviour of foreign high-technology corporations in Israel demonstrates typical location tendencies of high-technology enterprises. Intel, for example, opened its design centre in Haifa's science park in 1974. Major location factors were probably the availability of a suitable site in Israel's first science park and proximity to the Technion. In 1978, Intel set up a marketing and sales subsidiary in Tel Aviv's Atidim Science Park. This subsidiary, in charge of Israeli, Greek, and Turkish markets, was located in Israel's largest market and centre for international transactions. In 1985, Intel's largest facility in Israel, producing high quality and sophisticated silicon chips, was opened in Jerusalem, which enjoyed highest government incentives in a metropolis offering ample qualified labour.

Motorola located most of its facilities, including R&D, production, and marketing, in Tel Aviv; perhaps because its initial Israeli location decision was made in 1964, before the establishment of Israel's planned science parks and before the evolution of high-technology clusters in suburban locations such as Herzliyya. In 1987, Motorola established a new plant in the southern town of Arad, producing communications equipment for exports and portable cellular telephones for the Israeli market. Motorola followed the location strategy of earlier counterparts, searching for a location which was within the labour market of a metropolitan area, while enjoying preferred development zone incentives. After examining possibilities in cities like Jerusalem and Ashqelon, Motorola nearly settled for a site in the development town of Sederot, which is the nearest town to the Tel Aviv metropolis enjoying an 'A' development zone status (except for settlements in the occupied territories). However, the mayor of Arad in the Negev pressed for locating the plant in his town. Being closely affiliated with Yitzhak Rabin, then Minister of Defence, he managed to persuade Rabin to demand that the plant, which was relying on the local defence market, locate in Arad.

The case of Motorola indicates perhaps the beginning of a new stage in the process of dispersal of high-technology branch plants, this time to development towns which are not within commuting distance of metropolitan areas. Such plants have usually been engaged in simpler and relatively low-wage production processes, employing a high percentage of women. Two other examples of smaller high-technology plants in remote development towns have been Vishay Israel in Dimona in the Negev and Rada in Bet Shean in the eastern Galilee. The former has been engaged in routine operations, but to some extent the latter has been an exception. Rada was established in Haifa in 1970 as a spin-off from Elbit (a large-scale defence contractor) and specialised in computer systems for military aviation applications. In 1979, the growing plant moved to Bet Shean under the influence of the Ministry of Defence. Company headquarters and some of the R&D activities located in Herzliyya, and the firm also operated subsidiaries in the US. Nevertheless, the unique feature of Rada's plant in Bet Shean, which employed eighty workers in 1990, is that it is also engaged in R&D despite its remote location. Production in small-batch technologies made the complete separation of production and R&D unprofitable. Hence, despite difficulties in recruiting professional labour and the need to be self-sufficient to a large extent (due to the lack of local subcontractors and suppliers), the plant was able to operate successfully in Bet Shean. It was even successful in overcoming the blow of the cancellation of the Lavi fighter-plane project, for which it had won the contract for developing the computer.

IMPLICATIONS OF THE CRISIS OF THE 1980s – THE MIXED BLESSING OF UNSELECTIVE DISPERSAL

The crisis in the electronics industry since 1985 illustrates the fact that, in the long run, branch plants of Israeli and American high-technology corporations may not have the desired effect on development towns. The Elscint branch plants in Nazerat Illit and Jerusalem, for example, closed down soon after their parent company, based in Haifa, entered a phase of severe crisis and restructuring. As to foreign multinationals, while the government financed nearly all the investment in their high-risk production facilities, and offered additional benefits, it was rewarded only with some general promises by the corporations with regard to exports and employment, subject to external and internal conditions of the firms (State Comptroller of Israel 1984). Intel's plant in Jerusalem buys material inputs and know-how from its American parent corporation and sells the final product to the same parent corporation. Pricing of the product made in Jerusalem reflects, therefore, internal decisions of Intel, based on tax and other considerations, rather than market mechanisms (State Comptroller of Israel 1988).

The case of Xicor, a Silicon Valley-based micro-electronics corporation, demonstrates the problematic nature of investment by foreign high-technology corporations in Israel's development towns. In 1989, the corporation's ex-Israeli president negotiated with the government and with the towns of Beer Sheva and Karmiel in order to receive maximal incentives for establishing a large semiconductor production plant. The proposed investment was of $268 million and the plant was planned to employ up to 1,700 workers (Eshet 1989). Karmiel succeeded over Beer Sheva, perhaps due to its relative proximity to the Haifa metropolis and the prior existence of high-technology plants in the Karmiel area. However, the major negotiations were conducted with the central government. Xicor demanded benefits beyond those generally granted to firms in an 'A' development zone, basing its request on the precedent of benefits received by Intel's plant in Jerusalem. It has been argued that the terms agreed upon between Xicor and the government in 1989 were not beneficial to the Israeli economy (Eshet 1989; State Comptroller of Israel 1991). These arguments are based on comparing the cost of government incentives with benefits to the national economy – salaries paid to local workers, taxes, and inputs purchased locally minus their import-component. Of a predicted price of $3,600 for the final product, Xicor indicated that only $368 would be value added in the Israeli plant, all the rest being R&D, marketing, administration, interest, and profits paid mainly in the United States. Salaries in the Israeli plant were to be similar to the Israeli average and total annual salaries may not have reached the annual public subsidies. The amount of corporate taxes to be paid was difficult to predict, particularly since in multiplant enterprises the profits of specific plants can be manipulated.

Xicor's president admitted that its proposal might not be most attractive from the Israeli point of view. However, the new plant could contribute to the development of the semiconductor industry in Israel; besides, the proposal was the best offer Israel would get. In 1989, foreign direct investment in Israeli industry was minimal, and without the incentives the plant would have located in Scotland or Ireland where incentives are comparable and where there are no 'Intifada' or military reserve duties. Whereas the government approved a large-scale package of grants and other incentives for the proposed investment, doubts were raised as to whether Israel enjoyed any comparative advantage in the proposed operations based on high capital intensity and relatively cheap labour. In 1991, Xicor backed off from its Israeli venture, putting the blame on difficulties in obtaining permission to operate the plant on Saturdays. Instead of their original large-scale investment, they proposed a much more modest plan for establishing an Israeli design centre.

Just as Xicor was cancelling its plans, another large-scale investment proposal was looming on the horizon, this time from Atari. Atari's proposed semiconductor plant was to meet shortfalls created by the closure of an Atari factory in Taiwan, and to take advantage of the Israeli trade agreement with the EEC. Again, special benefits were negotiated with the government for the $225 million investment. The government favoured the proposed investment in principle, despite doubts as to whether Israel had a comparative advantage in the type of manufacturing that Atari was proposing. However, after several months in 1991, Atari, which was suffering from staggering losses, cancelled its plans, blaming the slow reaction of Israeli bureaucracy.

These two examples illustrate that the Israeli government was ready to allocate large-scale incentives to multinational foreign investment, despite the uncertain prospects of these investments for the Israeli economy. The government probably saw value in strengthening the local semiconductor design and production capability, even if the proposed investment would not necessarily be beneficial in the narrow sense to the Israeli economy. Indirect positive influences of such foreign investments could include employment and income multipliers, and upgrading technological personnel, as well as setting new standards of quality control.

It can be argued, however, that types of operations suiting Israel's comparative advantages are better demonstrated by companies such as DSP, which operates a design centre in the Tel Aviv metropolitan area employing eighty workers. Production in DSP is carried out by subcontractors abroad, particularly in the United States, taking advantage of the over-capacity in the global semiconductor industry, rather than trying to carry such operations in Israel. The company's head office is located in California, where the core of the industry and potential subcontractors are located, and the president has been living in Japan keeping close contacts with potential customers. However, the prospects of such design centres locating in peripheral areas are slim in view of the close integration with global production and marketing activities that are required. These operations do not require large capital investments, large tracts of land, or cheap, non-professional labour, and their dependency on international communications makes their dispersal out of the major metropolitan areas most unlikely.

Small software and other computer-related enterprises have been established in small non-metropolitan communities since the late 1970s. The late 1970s and the 1980s witnessed a process of dispersal of groups of people of higher economic strata into relatively small non-metropolitan localities. Some groups included people with political-ideological motivation who aspired to settle in the territories occupied by Israel since the 1967 war. Other groups were motivated by 'quality of life' factors, searching for low-density single-family housing. Two sectors which have gained ground in these settlements are small electronics and software enterprises. Of the sixty-four Jewish settlements in the occupied West Bank (excluding the Jordan Valley) as of spring 1983, thirteen had at least one firm pertaining to computer-related industries. Of nineteen new settlements in the Tefen and Segev regions of the Galilee, four had at least one such firm (Salomon and Razin 1985).

The geographical distribution of software and related industries is very centralised (see Table 13.1). However, interviewing owner-operators of some of the new peripheral software shops indicated that much of the professional work in this high-skill industry can be performed at any location, as the major necessary inputs are the computer and manpower. Thus, if the manpower exists, the only remaining problem is of communications. Here opportunities can be opened up by an improved telecommunications system (Salomon and Razin 1985). Computer-related industries have been candidates for relatively early adoption of telecommunications technologies. However, the telecommunications system of Israel did not provide equal service across geographical space. The inner parts of all three metropolitan areas enjoyed the best access to the telephone network. The outer fringes of the metropolitan areas did not have an advantage over the periphery, but they did enjoy an advantage by being able to contact more subscribers of the network at relatively low rates, and were able to use travel more cheaply as a substitute for telecommunications (Salomon and Razin 1988). The interviews with peripheral software shop operators suggested that the peripherally located enterprises can bear a high cost for maintaining contact with the market as long as the profit margins in the industry are wide enough (Salomon and Razin 1986). Thus, since the slow-down of the mid-1980s, as profit margins in computer-related businesses have narrowed substantially, the phenomenon of non-metropolitan electronics and software enterprises has ceased to expand significantly.

14

THE RE-EMERGENCE OF LOCAL DEVELOPMENT STRATEGIES

FACTORS ENCOURAGING THE EMERGENCE OF LOCAL DEVELOPMENT POLICIES IN ISRAEL

The increasing role of local authorities in public efforts to promote industrialisation have had a marked impact on shifts in Israel's industrial geography since the late 1970s. Several interrelated processes have encouraged the emergence of locally initiated economic development policies. External economic and political shifts were of prime importance. The economic crisis of the 1970s and 1980s and the declining power of the central government have both worked to decrease the effectiveness of the national spatial industrialisation policy, which has failed to adjust to economic restructuring processes. Increasing pressures on welfare state mechanisms have led to cuts in public budgets available for such policies, and although these policies have generally been continued, their effectiveness has diminished.

The political system shaped by the Labour party, in power until 1977, was characterised by high levels of centralisation and government intervention in the economy, due to the party's socialist ideology and to the strong charismatic personalities of its leaders. Urban local authorities were in a weak position, suffering from being considered strongholds of the Right. However, the decline in authority of the central government, caused by the failure to anticipate the 1973 War and by a wave of corruption scandals, which hastened the exit of the old charismatic leadership, created new opportunities for local initiative. The right-wing Likud party, which took power in 1977, was officially committed to *laissez-faire* ideology. The municipal electoral reform, which led to the direct election of mayors in 1978, and the budgetary policy of the 1980s, which increased the reliance of local authorities on municipal taxes, contributed to a shift in the balance of power between local and central government (Elazar and Kalchheim 1988). Thus, the economic crises and political realities of Israel during the 1980s reflected the increasingly fragmented action of central government. This also allowed for local initiative after about twenty years in which formal criteria and dominant action by the central government served to inhibit such strategy.

Increasing political mobilisation amongst immigrants of Asian and African origins, who formed the overwhelming majority in development towns, had a particular role in the emergence of a generation of more potent and influential local leaders. Local residents of Asian and African origins, who served in the past as mayors of development towns, depended on the bureaucracy of nation-wide political parties (Weiss 1970; Cohen 1974; Elazar and Kalchheim 1988). During the 1970s and 1980s, a new generation of political leaders rose to power in development towns. They were less dependent on party 'bosses', and were mostly affiliated with the right-wing Likud party. Their rise to prominence was linked to the emergence of protest by Jews of Asian and African origins against patronage of the Labour party (Gradus 1984). Young mayors of some development towns became more active in determining the future of their towns, utilising their power positions in the national political arena to promote local development.

Local initiative was also encouraged by the changing planning perceptions, which emphasised mobilisation of indigenous resources. As already discussed, Israeli regional economic policy tended to adopt strategies of leading countries, usually within a time lag of about a decade. Thus, awareness of the significance of local ownership and entrepreneurship surfaced in Israel during the 1980s (Avraham 1985; Razin 1988b), stimulated by the British and American planning literature focusing on these issues. The renewed emphasis on the involvement of local authorities in economic development efforts in Israel was also influenced by the re-emergence of municipal boosterism as a common practice in the Western world during the last two decades (Boyne 1988; Harvey 1989; Campbell 1990).

COMPETING STRATEGIES

Emerging local development efforts in Israeli non-metropolitan towns during the 1980s consisted primarily of action taken by individual mayors in negotiating with government agencies over the allocation of the meagre public resources or in mediating between potential investors and the political-bureaucratic establishment. Local leaders have also been engaged in marketing their towns and supporting the existing local economic infrastructure. Municipal development corporations have also been set up. These have initiated economic projects, gaining increasing popularity due to their organisational flexibility and ability to circumvent legal and administrative barriers in the operation of the municipalities themselves (Klausner and Shamir-Shinan 1988).

The specific development strategies instigated by each town have depended on its location, site, and other attributes, as well as on the power position and personality of its town leaders. Nevertheless, strategies undertaken by non-metropolitan towns can be classified into three major groups: (1) effective exploitation of the capital incentives of the national spatial industrialisation policy; (2) utilisation of opportunities created by proximity to a metropolis; and (3) a search for development alternatives, particularly when a town is in an inferior position with respect to the other strategies.

Utilisation of spatial industrialisation policy measures

The most common local development strategies have focused primarily on improving access of towns to the national government incentives system. Thus, mayors have pressed for receiving higher development zone incentives, offered additional benefits, advertised their towns as attractive business environments, given potential investors personal treatment, and appointed economic development officers in an effort to by-pass bureaucratic procedures associated with receiving government assistance (Klausner and Shamir-Shinan 1988).

Towns enjoying most favourable locations on the map of assisted areas were most successful in this strategy, Migdal HaEmeq and Sederot being two of the most notable examples (see Figure 11.1d). During the late 1970s, Migdal HaEmeq had a stagnant industrial base dominated by a few large food, textile and leather plants. The turn-around occurred during the early 1980s, largely as a consequence of a change in town leadership, with renewed industrial growth accompanied by diversification and emergence of new industries, particularly in electronics and metal products. The dynamic mayor elected in 1978 represented the Likud party, which had won the national elections in 1977. He successfully pressed for the upgrading of his town's status from a 'B' to an 'A' development zone. This change, which took effect in 1978, was considered temporary, but has in fact endured to the present (1992). It gave Migdal HaEmeq an immense advantage, since it is the closest town to the Haifa metropolis enjoying 'A' development zone benefits (see Figure 11.1d). Thus, for example, Migdal HaEmeq was 'sold' to National Semiconductor as a site within convenient commuting distance of the Haifa metropolis which also enjoys the highest development zone benefits. Migdal HaEmeq's mayor was most active in channelling public budgets to his town and attracting industries, as well as in creating a positive industrial climate and an improved town image (Schwartz 1988).

Similar processes were occurring in other development towns located within commuting distance of the Haifa metropolitan area, such as Maalot, Karmiel, and Afula. A major element in most strategies was to concentrate efforts on attracting a large high-technology plant, which was intended to serve as an 'anchor' plant (Schwartz 1988). This plant was usually a branch plant of a firm having a central facility in the Tel Aviv or Haifa metropolitan areas. Such a plant was intended to provide better-paid jobs, to change the industrial image of the town, to reduce uncertainties for other plants about the operation of advanced industries in the specific location, and gradually to contribute towards the formation of pools of local qualified labour and information networks.

In southern Israel, Sederot is the nearest town to the Tel Aviv metropolitan area having an 'A' development zone status (see Figure 11.1b–e), with the exception of settlements in the occupied territories east of Tel Aviv. Thus, it offered maximal incentives with minimal uncertainties as to labour supply and business contacts relative to other towns receiving equal incentives. Active political leadership has also exploited these advantages since 1983; hence, Sederot has had more industrial floor space built during the 1980s than any of the other towns in the region (Razin 1990b).

Qiryat Gat, which is more centrally located than Sederot, is an example of a less successful economic development path. Qiryat Gat has a 'B' development zone status and has suffered from a location inferior to that of Qiryat Malakhi, which is entitled to similar incentives. Hence, despite the efforts of its local Economic Development Corporation, it has been largely unsuccessful in changing its industrial character and reducing its dependence on the large, vertically integrated Polgat textile complex. This complex employed between 4,000 and 5,000 workers during the early 1980s, but during the late 1980s it entered a phase of contraction and restructuring. The number of its employees in Qiryat Gat declined to about 3,000, and in 1991 the complex was still losing money and still contracting. Polgat had a major role in the development of Qiryat Gat's economy during the 1960s and perhaps also during the 1970s. However, the dependence of the town on Polgat became increasingly problematic during the 1980s. A significant number of the jobs offered by the complex were no longer attractive to local residents. Thus, Arabs from the occupied territories accounted for a large proportion of production workers – as high as 50 per cent in the main textile plant of the complex, which also employed 120 foreign workers from Portugal (Felsenstein and Schwartz 1991). The contractions in Polgat were not compensated by job formation in other enterprises, and most of those laid off had to look for new jobs out of town. Thus, the town, which was one of the better planned and more successful development towns during the 1960s, has entered a period of demographic and economic stagnation (Razin 1990b).

The experience of towns basing development efforts on exploiting government assistance indicates that, in an era of economic stagnation and contracting public budgets, such policies cannot suffice to significantly change the prospects of more than just a few towns. Rather, instead of increasing the limited pool of industries willing to locate in these towns, such policies may have caused the diversion of resources from one town to another. Only towns offering the combination of maximal incentives and most favourable location, enhanced by active and effective leadership, were able to succeed in attracting industries. The success of Sederot might have been at the expense of other towns located farther south, such as Ofaqim (Bar-El and Schwartz 1985), which are entitled to the same level of incentives but suffer from inferior locations. It has also been at the expense of the larger town of Ashqelon, 15 km closer to Tel Aviv but lacking any development zone status. Similarly, the success of Migdal HaEmeq has been at the expense of towns to its east.

In the few advantageous towns – Sederot, Qiryat Malakhi, and a ribbon of development towns near the fringes of the Haifa metropolitan area – industrial promotion strategies have been successful in changing the economic and social atmosphere of previously backward and stagnating towns. However, industrial job creation in the periphery alone has not created a self-sustaining development process. New branch plants in advanced industries tend to create jobs that are prone to lay-offs at periods of economic stagnation and corporate restructuring. In addition, the 'leading' high-technology branch plants have not usually met (often inflated) growth prospects, and Israeli-owned corporations particularly tended to close down their peripheral branch plants during the marked slowdown in the electronics industry that began in 1985.

Furthermore, industrialisation efforts have not had a marked influence on unemployment levels in the development towns, which have remained relatively high in Sederot and Migdal HaEmeq, the two Israeli archetypes of the successful implementation of such industrialisation policy during the 1980s (Lavy 1988; Schwartz and Felsenstein 1988). A high proportion of employees in these new industries have been non-locals, since industries were initially attracted to these towns due to the ability to draw upon metropolitan labour markets while enjoying development zone benefits. It could also be that the existence of more job opportunities in the successful development towns might have encouraged a greater level of participation in the labour force by subsectors of the population, such as housewives, that had previously stayed at home and had not appeared as unemployed.

Felsenstein and Schwartz (1991) identified another specific difficulty in towns that were successful in attracting industries due to the combination of development zone incentives and proximity to a metropolis. Plants attracted to such towns have tended to be those types having intense links with metropolitan regions. Hence, location in the development town creates difficulties for them, since they suffer from disadvantageous location relative to their competitors who have stayed in the centre. Plants in the more peripheral regions have tended to be engaged in more routine activities which did not require intensive linkages with the metropolis and thus have complained somewhat less of disadvantages associated with their remoteness.

In conclusion, industrialisation has taken a much less uniform pattern during the late 1970s and the 1980s than it did in the preceding decades, expanding only at select, profitable locations. Nevertheless, these few poles of industrial growth are still insufficient for providing solutions for long-term unemployment problems, or for the generation of self-sustained growth. These poles are mainly composed of branch plants, and are not located in towns that are sufficiently large to permit the replication of growth-pole strategy. Concentrating efforts on utilising the government's incentives as an economic development strategy has been rooted in perceptions and patterns of action that have assumed such strategies as most effective in creating large numbers of jobs in a short time. However, the limitations of this strategy suggest that it can only be justified in those towns enjoying a particularly advantageous location on the map of the development zones.

Taking advantage of proximity to a metropolitan area

Exploiting advantages associated with proximity to a metropolis by offering attractive housing opportunities and promoting economic integration with the metropolis has been the local development strategy which has achieved the most visible results. The role of local initiative is crucial in shaping such a local development path, since the incorporation of a town into a metropolitan system may take many forms. Such strategies have consisted of promoting the development of high-standard residential neighbourhoods for commuters and of economic activities such as space-consuming industries and back-offices. Migdal HaEmeq has been a clear example of a town which focused efforts primarily on suburbanisation of economic activities from the Haifa metropolis. In contrast, Yavne, located near the southern fringe of the Tel Aviv metropolitan area, provided the best example of local initiative in formulating a suburbanisation strategy. Yavne was a stagnating backward city until the mid-1970s. Its mayor, Meir Shitrit, elected in 1974, opted for developing it as a low-density satellite of the Tel Aviv metropolis. Yavne was among the forerunners in initiating a project to provide single family dwellings for medium-income households. Its unprecedented success was inspired by a correct reading of the changing housing preferences in Israel, and opened a new range of opportunities for the city. The housing project was followed by efficient up-grading of the school system and by attracting a large group of army officers and their families as a core of households ready to build their houses in the once-depressed immigrant town. The growth process relied upon ease of commuting to the Tel Aviv metropolis.

This rapid transformation has not been without problems, such as social polarisation and rising land and housing prices. At the onset of the 1990s, Yavne has been pursuing a more balanced development path, re-emphasising its advantages for industries discouraged by high rents in the major industrial areas of the Tel Aviv metropolis. Nevertheless, the change effected in Yavne since the early 1970s has been clearly a function of an effective local policy, which materialised despite the lack of a significant industrial development component.

The search for alternatives

The majority of development towns have lacked the advantageous locations needed to succeed in the strategies mentioned above and have thus been pursuing alternative paths of development. This has been the case for most of the towns in the remote periphery, and also for towns in semi-peripheral regions that suffer from intra-regional disadvantages. Ashqelon is a prominent example of such a town. It is too remote to base development efforts on proximity to Tel Aviv (only 5 per cent of its working population commuted to the Tel Aviv metropolis in 1983), and is in a particularly weak position to attract industries. Ashqelon is surrounded to the east and south by towns enjoying superior 'B' and 'A' zone status (see Figure 11.1b–d). Thus, its industrial base, formed during the 1950s and 1960s when the town did enjoy a privileged development zone status, has hardly grown or restructured.

The city's political leaders, elected in 1978, acknowledged Ashqelon's disadvantage in attracting industries and focused efforts on obtaining a temporary 'A' development zone status for high-technology plants interested in locating in its southern industrial area. These attempts failed because the Ministry of Finance opposed them due to their possible negative impact on the government budget. Poor relations between Ashqelon's mayor and the Minister of Industry and Trade may have also hampered prospects for promoting Ashqelon's case. The efforts of Ashqelon's mayor to be elected to Israel's Parliament (Knesset), finally successful in 1988, were motivated by his perception that the attitude towards Ashqelon would not change without the city being represented at the national centres of power. Indeed, Ashqelon finally secured a temporary 'B' development zone status in 1990.

Facing difficulties in industrialisation, local authorities have frequently pursued tourism as an alternative. However, the large-scale potential in tourism is limited to a few locations in Israel. Promoting local entrepreneurship is another economic development strategy often advocated by external researchers and consultants (see Chapter 15). Direct support systems for small businesses may be promoted by municipalities in cooperation with local associations of small businesses or various public and non-profit agencies. The construction of multi-purpose buildings for small industries has also been perceived by municipalities as an attractive option, although usually as an uneconomical one due to low market rental rates of the 1980s. However, the need for removing bureaucratic barriers for the operation of small businesses, revealed to be of major importance abroad (Maillat 1988) and in Ashqelon (Razin 1990b), has encountered difficulties.

Other options have been to promote central services and to improve the attractiveness of the local housing market (Razin 1990b). An additional attempt has been made to identify the potential in promoting producer services in development towns (Schwartz and Felsenstein 1988). Prospects for dispersing business services were found to be limited, due both to their strong concentration tendencies and the relatively small distances in Israel. This lack of friction of distance allowed for the purchase of top-level services in metropolitan areas by plants in the periphery. Some potential for dispersal was found in low-level producer services such as catering and transportation. Hence, these strategies have evolved, to a large extent, as 'no choice' alternatives of towns lacking an advantage with respect to incentives for industrialisation and proximity to a metropolis. It is unclear whether such alternatives, even if cost-effective in generating jobs within the limited confines of their resources, can produce more than marginal change in long-term economic trends.

THE SPATIAL IMPLICATIONS

The emergence of locally initiated development strategies led to the intensification of competition between towns during the 1900s. The Israeli urban system in non-metropolitan regions was initially planned on central place principles, and was greatly influenced for three decades by the government's spatial industrialisation policy. It can be argued that the shift from nationally directed industrial dispersal efforts to greater emphasis on locally initiated development strategies leads to the restructuring of the urban system by disrupting spatial processes that have occurred since the 1950s.

At the national level, the gap between core metropolitan regions and the remote periphery has widened. Semi-peripheral regions are also attaining a growing advantage over the remote periphery. Semi-peripheral regions are non-metropolitan regions whose urban evolution resembles that of the remote periphery, but which have been gradually entering the metropolitan sphere of influence with regard to commuting patterns and to the location of economic activity. During the 1980s, the Israeli semi-periphery included the southern coastal plain, south of the Tel Aviv metropolitan area; the western Galilee, east of the Haifa metropolitan area; and parts of the occupied territories, in the vicinity of the Tel Aviv and Jerusalem metropolitan areas (see Figure 14.1). These parts of the occupied territories differ from the other semi-peripheral regions in their settlement evolution. Also, the Arab majority living in these areas face a different spatial core-periphery structure than the Jewish minority, to whom Figure 14.1 refers.

The semi-periphery has enjoyed relative proximity to major political and economic centres and thus has had better opportunities to succeed in locally initiated development efforts. Hence, the emerging pressure group representing development towns has been led by personalities from the semi-periphery who have an interest in preserving development zone benefits for their towns. This situation has given semi-peripheral towns an immense advantage over remote towns.

Since the 1970s, the towns in Israel's semi-periphery, particularly those in the southern coastal plain, have come to be a major venue for the emergence of a new, young, local, political leadership in the development towns. Towns in Israel's semi-peripheral regions were in an extremely influential position in the national political arena of the late 1980s. Many were represented in the Knesset by personalities whose power positions were largely secured through successful careers as mayors of their towns. These leaders could represent the periphery while enjoying relative geographical proximity to political power centres and having ample opportunities to achieve progress in their towns in comparison with more remote towns. Political activists coming from more remote development towns, such as Dimona and Bet Shean, have been less likely to follow the same mobility path, since they have lacked similar opportunities to succeed as mayors.

Among semi-peripheral regions, intraregional competition has intensified as towns follow diversified paths of development.

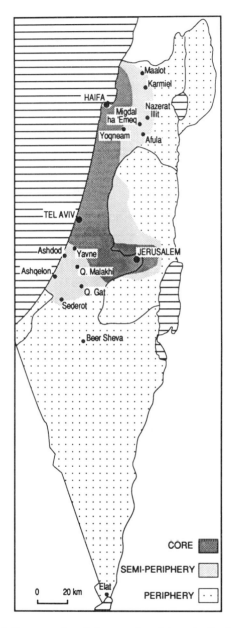

Figure 14.1 Core, semi-periphery and periphery in the Israel of the 1980s – a preliminary sketch

During the 1950s and 1960s, there was little to distinguish between the local economies of the various development towns, since the government regulated the industrialisation process, distributing the dispersing industries among development towns. Economic investment patterns were therefore similar in most development towns. During the 1980s, however, advantages and disadvantages at the micro level have gained particular importance in the semi-periphery, and towns as close as 15 km apart can be destined for very different development paths. This transformation supports the argument of Smith and Dennis (1987) that equalisation of conditions dominates at the end of long cycles of expansion, whereas a more fragmented intraregional pattern emerges as crises begin and deepen. Thus, it can be argued that the geographical scale at which economic variations are to be analysed has been substantially reduced.

15

THE LOCAL ENTREPRENEURSHIP OPTION

INTRODUCTION

The emerging world-wide interest in the role of small businesses in regional economic development has not by-passed Israel, and the small business sector is expected to become an important factor influencing Israel's industrial geography during the 1990s. Local entrepreneurship has been increasingly advocated by planners and researchers as a remedy for economic difficulties in Israel's development towns (Avraham 1985). The calls for promoting local entrepreneurship, however, were ignored for a long time by policy-makers. In addition to being contrary to prevailing ideological beliefs, promoting entrepreneurship did not have the same appeal of offering a highly visible and quick influence on local economies as did the effect of attracting large, externally owned plants or projects aimed at the tourism industry. Shrinking opportunities for attracting new, externally owned plants and limited potential in tourism, however, have led policy-makers to develop an interest in local entrepreneurship strategies. Nevertheless, formation of instruments for promoting entrepreneurship by the government, local authorities, and other public organisations reached the 'take-off' stage only when the new wave of immigrants from the former USSR, which commenced late in 1989, created the urgent need to exploit all possible means for generating new jobs in the Israeli economy. While the regional dimension in these new instruments is explicit, their regional impact is far from certain, since, unlike the case of capital-intensive, mass-production plants, dispersal of a wide array of small enterprises may contradict locational market mechanisms.

TRENDS AND SPATIAL VARIATIONS

The Israeli small business sector stagnated during the period 1961–83 and the percentage of self-employed among the economically active urban Jewish population constantly declined (see Table 15.1). The decline was mainly evident among the self-employed, who had no paid help, rather than among those owning and managing larger businesses. The Tel Aviv metropolis had the highest rates of self-employment, whereas the development towns, both centrally and peripherally located, had the lowest (see Table 15.1). Apparently, the large manufacturing sector in most development towns, consisting mainly of mass-producing and externally owned plants, did not promote many opportunities for self-employment and had a negative impact on local entrepreneurship (Razin 1990a).

All three metropolitan areas enjoyed a large and increasing advantage over development towns with respect to self-employment opportunities in business and other professional services. The Tel Aviv metropolis enjoyed a particular advantage in manufacturing and wholesale trade. Tel Aviv retained its role as a centre of small-scale industries having complex linkage patterns, such as diamonds and clothing. Diamond-polishing plants clustered in the vicinity of the diamond exchange in Ramat Gan (see Chapter 6), and small-scale garment manufacturing clustered around the major trade centre in southern Tel Aviv. The concentration of self-employment in the fashion-oriented clothing industry in the Tel Aviv metropolis even increased between 1972 and 1983 from 67.2 per cent to 77 per cent. The decline of self-employment in manufacturing was most marked in the secondary metropolitan areas of Haifa and Jerusalem (see Table 15.1), where the nation-wide trend of industrial dispersal was not compensated for by thriving complexes of small manufacturing plants enjoying agglomeration economies.

Over the period 1961–83, a shift can also be discerned in the sectoral composition of self-employment. In metropolitan areas the major trend among the self-employed was from retail to various professional services, while in development towns the share of construction contractors, carpenters, metal workshops, car repair shops, and some retail activities, showed the highest growth. These were largely non-growth blue-collar occupations. The share of manufacturing among the self-employed in development towns was relatively small, although in 1983 it was no longer below that of the Haifa and Jerusalem metropolitan areas. The major and only

Table 15.1 Israel's urban population – self-employed, and self-employed in manufacturing, among Jews by location, 1961–83[a]

Location	Self-employed (%)			Self-employed with 3+ salaried workers (%)[b]			Self-employed in manufacturing[c]		
	1961	1972	1983	1961	1972	1983	1961	1972	1983
Tel Aviv Metropolis	20.5 (64.5)	17.7 (90.0)	13.3 (113.0)	1.6	2.3	2.0	28.0	23.9	21.4
Haifa Metropolis	16.7 (18.0)	12.9 (24.0)	10.1 (28.0)	1.3	2.1	1.8	25.0	19.0	16.6
Jerusalem Metropolis	13.2 (11.7)	11.5 (16.1)	10.4 (22.6)	1.0	1.7	1.5	24.9	17.6	13.0
Veteran towns in the Coastal Plain	18.1 (6.0)	16.6 (9.9)	12.2 (13.3)	1.4	2.0	1.7	24.3	17.1	17.6
Development towns in the Coastal Plain	12.9 (2.8)	10.1 (8.3)	8.3 (12.2)	0.5	0.9	0.8	18.4	16.2	17.7
Development towns in the periphery	10.7 (5.0)	9.4 (12.1)	8.3 (18.3)	0.5	0.9	1.0	18.5	16.8	14.9
Beer Sheva	13.2 (2.6)	11.8 (5.1)	8.9 (7.6)	0.9	1.0	1.3	17.2	17.8	15.0
Total	18.2 (110.0)	14.9 (165.4)	11.6 (215.1)	1.4	2.0	1.7	26.4	21.5	19.1

Source: Razin (1990a), based on Central Bureau of Statistics, *Census of Population and Housing*, 1961, 1972, 1983.

Notes:
[a] Total working population in thousands in parentheses (20 per cent sample of the census).
[b] The self-employed with 3+ salaried workers are part of a wider category of self-employed. The figures relate to percentage of self-employed with 3+ employees out of the total working poulation in each location.
[c] Percentage of self-employment in manufacturing out of the total number of self-employed in each location.

significant variable influencing self-employment in manufacturing in the development towns was distance from the nearest metropolitan centre. A remote location significantly decreased opportunities for small manufacturing establishments (Razin 1990a). Self-employment opportunities in retail and construction which were not confined by distance were restricted by the small size of the local market and the limited purchasing power in the development towns.

SHIFTING ATTITUDES TOWARDS ENTREPRENEURSHIP – FROM AN EVIL TO A BLESSING

The resurgence of the small-business economy in many Western countries from the mid-1970s onwards (Storey 1988; Brock and Evans 1989) did not induce a similar trend in Israel for more than a decade. The Israeli political-economic system was not very receptive towards small entrepreneurs due to an historical labour-socialist tradition that engendered deep antagonism, and at certain periods even hostility, towards the self-employed sector. Small-business owners were perceived as unproductive middlemen who lived by exploiting industrial and agricultural workers. Calls for public policies concerning the small business sector were frequently focused on the need to reduce alleged tax evasion by the self-employed, which, it was considered, gave them an unjust advantage over salaried employees. The drive towards self-employment in Israel was probably also delayed because economic crisis conditions were absorbed between 1973 and the mid-1980s by rapid inflation and an accumulating foreign debt, while massive unemployment was avoided.

Greenwood (1990) emphasises the unfavourable climate for entrepreneurship which he claims has made Israel one of the few non-Communist countries where Jews did not gravitate to small businesses. Among the 'nightmares' experienced by Israeli small businesses, one can include the severe consequences of annual military reserve service on sole proprietor businesses and the self-employed, inferior treatment by welfare state mechanisms, the government's role in the capital markets which dried out most traditional sources of financing for small businesses, difficult access to vital information on procedures required for establishing a business, and all the other evils of the bureaucratic treadmill which discriminate against small businesses in favour of large private and public corporations and non-profit organisations. These obstacles have been magnified by common perceptions, which include a mythical fear of taxmen, licensing, and unexpected discouraging economic or political news.

However, a gradual change in attitude in favour of the small business sector has occurred in Israel. This change has been provoked in part by a political-ideological shift. The right-wing Likud party, which assumed power in 1977, was at least officially more committed to free-enterprise ideology than the Labour party, which had led Zionist and Israeli politics until then. Pressures originating from infiltration of 'new right' ideologies from Britain and the United States were perhaps of even greater impact. The change was also a consequence of new economic realities of stagnation and crisis in many of Israel's largest industrial corporations and public organisations during the late 1980s, as well as of rising unemployment and increasing pressures on the government's budget. These conditions drove various groups, such as retired army personnel, into self-employment.

Diffusion of planning perceptions and practices from Western countries had an additional role in inducing change in attitudes toward entrepreneurship. Advocates of small business-oriented policies in Israel (Avraham 1985; Razin 1988b; Greenwood 1990) have been guided by the revival of interest in the role of small firms in job creation and economic development in European and North American economic planning circles (Birch 1987; Giaoutzi *et al.* 1988; Piore and Sabel 1984). Calls for emphasising small business formation, while grounded in recent professional literature and experience of Western countries, had roots in two very different ideologies. On the one hand, many proponents of this approach viewed the promotion of the small business economy in the context of conservative free-market ideology. Thus, promoting entrepreneurship has been perceived as part of a wider action to enhance competition, reduce bureaucratic red tape, and remove rigidities in the labour market by cutting down the strength of labour unions. Other proponents, conversely, have their roots in a liberal-socialist ideology of development from below (Stöhr 1981). These proponents have stressed the need for local communities to have greater control over their economic fate, and to reduce their dependence on externally controlled firms (Watts 1981; Razin 1988b) and externally operated development policies. Hence, promoting entrepreneurship has been considered as a means of reducing interregional and interethnic inequalities, and not only as a tool to enhance the efficiency of market mechanisms. Another position is to view entrepreneurship as no more than a no-choice alternative. For example, an industrialisation strategy based on local entrepreneurship has been proposed for rural settlements where very few location advantages for external firms exist (Bar-El and Felsenstein 1990).

The power accumulated by the Israeli Association of Chambers of Commerce during the 1980s demonstrated the shift in the position of the small business sector in Israel's political arena. Led by the dynamic Danny Gillerman, the association, representing mainly small-scale non-manufacturing enterprises, succeeded in securing a role in the top-level economic policy forum that coordinates policy between the major economic forces at the national level. These coordination efforts were previously reserved for the government, the Federation of Labour, and the Israeli Association of Manufacturers, none of which represented interests of manufacturing and non-manufacturing small businesses. However, the political-ideological shift, combined with the calls to promote entrepreneurship, was unable to initiate the large-scale public action required to create the instruments necessary for promoting successful entrepreneurship. Only the immense economic pressures associated with the mass immigration from the former USSR have created the sufficient preconditions for such a policy shift, largely as a no-choice alternative.

PROMOTING ENTREPRENEURSHIP AND THE DEVELOPMENT OF THE PERIPHERY – EARLY ATTEMPTS

Consolidation of the positive attitudes towards entrepreneurship was slow, and only a few pioneering efforts for promoting entrepreneurship were implemented during the 1980s. Three 'early birds' for entrepreneurship-oriented regional development strategies were the industrial village initiative, the Ganei Taassiya incubator facility, and loan funds for small businesses initiated in two development towns by the Jewish Agency's Project Renewal.

The concept of the industrial village was proposed by the Jewish Agency in 1973. It had very little in common with the origins and goals of subsequent entrepreneurship-oriented policies. Rather, it was perceived by the Jewish Agency as a means to establish new rural settlements, despite the exhaustion of the supply of agricultural land needed to support them (Applebaum and Newman 1989). Implementation was mainly in the Segev region in the western

Galilee, east of Haifa. The plans to establish a cluster of industrial villages in the Segev region were formulated in 1976, with the motivation of forming a Jewish presence in an area which was predominantly Arab. The villages were to be organised as cooperative moshavim (see Chapter 18), but over time most of them were transformed into the Yishuv Kehillati form of settlement (Newman 1984), which depended more on private forms of entrepreneurship.

Of several industrial villages established in the western Galilee, only two – Yaad and Manof – have achieved considerable success. The whole idea proved to be problematic. First, the diversified labour needs of manufacturing plants, which are typically characterised by clear hierarchical structure, presented a problem, particularly in villages wishing to retain some elements of a cooperative structure. Moreover, the initial capital required for establishing a small industrial plant was greater than the capital needed for developing an agricultural farm, for which the level of support by the Jewish Agency was geared (Applebaum and Newman 1989). Finally, most settlers lacked entrepreneurial motivation. They aspired to live in rural surroundings and accepted the necessity to commute to the Haifa metropolis or to other large employers in the western Galilee. The two successful industrial villages were characterised by highly coherent and homogeneous groups, with a common background and a clear vision of their future joint venture. The first – Yaad – consisted of graduates from the Haifa-based Technion Institute of Science and the second – Manof – of South African immigrants. In all other cases, the industrial villages faced social and economic difficulties. Some industrialisation in new non-agricultural settlements has continued, but the industrial village form of settlement has generally ceased to grow.

A first explicit attempt to assist innovative entrepreneurship in general, and in the periphery in particular, was the Tefen industrial incubator park facility discussed in Chapter 13. This pioneering project has served as an example for later imitations. The Tefen incubator project served for several years as the sole example of its type in Israel, inspiring the imagination of various planners and policy-makers, who incorporated such projects in various local development plans. However, the motivation and skills for establishing such incubator facilities was still in short supply, and these proposals were little more than general slogans throughout the 1980s.

The Tefen initiative was accompanied by a massive public

relations campaign. It was introduced as a new form of productive Zionism. The founder claimed that it presented the antithesis of traditional deep intervention in the economy, as well as an alternative to the existing practices of Israeli labour unions, which are said to have perpetuated the dependency of voters on the existing establishment and to have blocked individually based development focusing on exporting industries. The Tefen initiative was promoted by slogans such as: 'The park was established in the remote Tefen area far from the eyes of the hostile government, in the same manner that the founders of the first kibbutz established their settlement in 1909 far from the older settlements which were dominated for a long time by representatives of the Baron Rothschild.' However, for all this rhetoric, the project and its tenants still were not opposed to receiving the highest development zone status, being located in an 'A' development zone. Thus, the Ganei Taassiya project can be viewed as a sort of private–public initiative, with the private actor providing most of the initiative and know-how and the government being the source of most of the financing and other benefits provided to peripheral industrial parks.

The Jewish Agency's Project Renewal – a large-scale project begun in the late 1970s to revitalise Israel's deprived neighbourhoods – was the first to implement a local entrepreneurship-oriented policy in development towns. Project Renewal has been engaged since its inception with housing, infrastructure improvement, and social services (Alterman 1988). Over time, it became evident that neglect in the sphere of economic development reduced the effectiveness of the project to transform long-term trends in neighbourhoods located in non-central towns. Consequently, a strategy for promoting local entrepreneurship was proposed as the most promising and suitable way to supplement the project (Avraham 1985). This new line of action also seemed attractive to Project Renewal officials as a strategy to secure a new role for the organisation, since their role as instigators of physical renewal and social services was near completion. Thus, in 1986 Project Renewal initiated loan funds for small businesses in two development towns (Qiryat Shemona and Ofaqim) (Klausner and Shamir-Shinan 1988; Felsenstein *et al.* 1991).

A major difficulty in most development towns has been the prevailing atmosphere of relying on the government for the provision of jobs (Shinan-Shamir 1984). Traditions of enterprise in Western terms that facilitate the formation of entrepreneurial economies, similar to those utilised as an upward mobility route by various immigrant groups in the United States (Portes and Bach 1985), or to those flourishing in 'Third Italy' (Bamford 1987), have not existed in the development towns.

Paths to entrepreneurship in development towns have been characterised by a high dependency on kinship and social networks for advancement through self-employment in blue-collar and distribution occupations (Razin 1990c). Entrepreneurs in development towns have increasingly depended on narrow family and social networks, whereas those in metropolitan areas have enjoyed better opportunities to gain relevant experience as salaried employees, as well as opportunities to cultivate useful business contacts and a broader range of business experience. Entrepreneurs in development towns have also had inferior professional qualifications and financial resources. Small businesses in development towns depended on suppliers of material inputs and services from the Tel Aviv metropolis. However, they have rarely been able to penetrate the metropolitan market with their products, since they have tended to operate with obsolete production methods and to suffer from information gaps, which have reduced their competitiveness (Felsenstein and Schwartz 1991).

THE TAKE-OFF – A MAJOR ROUTE FOR ABSORBING IMMIGRANTS OR A LATE ADAPTATION OF THE POLICY OF THE 1980s?

Early efforts to promote entrepreneurship were triggered by shrinking opportunities offered by alternative economic development strategies (Razin 1990b). However, the immense pressures created by large-scale immigration of Jews from the former USSR since 1989 has led policy-makers to grasp any opportunity for job creation which does not place a heavy burden on the public budget. Numerous schemes have subsequently been instituted that have been aimed at assisting entrepreneurial activity and small business. Extra-governmental public organisations such as the Jewish Agency and the Israeli office of the American Jewish Joint Distribution Committee have led the initiative, mobilising local authorities and other organisations as partners. These organisations and others, such as institutes of higher learning, have made joint efforts to establish technological and business incubators geared to new immigrants, loan funds for small businesses, and support centres for entrepreneurship.

Central government has also become directly involved in promoting entrepreneurship through two major instruments. First, a National Small Business Loan Fund for establishments of up to forty employees and a scheme for small business loan guarantees have been capitalised. Second, government has provided financial assistance for the technological incubator projects.

This emergent national economic strategy for promoting small business formation can have the effect of exerting a negative influence on already existing disparities between development towns and metropolitan areas. The major beneficiary of these non-spatial policies for promoting new business formation is likely to be the Tel Aviv metropolis, which enjoys the most diversified range of opportunities in the small business economy. Whereas government schemes do not include an explicit regional criteria for granting aid, those operated by the quasi-governmental agencies, which are often supported by government, have a clear regional element often targeted at particular towns or regions. Targeted programmes for business support and consultancy (one-stop business authorities) exist in metropolitan centres and those cities which have absorbed a large mass of immigrants from the former USSR (Beer Sheva, Ashdod and Netanya); and in two smaller localities in the western Galilee.

It is too early to present a detailed account and evaluation of these initiatives. Nevertheless, it seems that, while most schemes are oriented towards peripheral localities, it is those centrally located areas that are likely to become most active and have the most pronounced impact. Schemes in remote development towns may suffer from inferior opportunities in distributive activities, as well as from less facilitating environments for technological entrepreneurship.

The formation of new schemes to assist entrepreneurship raises several additional questions, including: (a) What is the effectiveness of public measures in promoting entrepreneurship? and (b) What is the specific entrepreneurial potential among immigrants in Israel? As to the first question, entrepreneurial development policies dealing with access to capital, management skills, information on supplier and buyer markets, and the political climate facing small firms can be cost effective in generating jobs within the limited confines of their resources. However, these policy tools of themselves are insufficient for creating the environmental conditions that characterise communities of small firms, such as the presence of an entrepreneurial culture and information networks. The successful development of the small firm economy in Third Italy, for example, can only be understood in the context of the particular social structure of the area, and not by a correct policy mix (Bamford 1987). Hence, such policies are likely to have an effect mainly in those places where entrepreneurship is already rooted (Mokry 1988). They can hardly be expected to produce more than marginal changes in trends influenced by broad long-term economic, social, and geographic processes.

To answer the second question, one has to examine specific immigrant communities, as well as the absorbing environment. It is unlikely that immigrants from the former USSR will be able to rely on intensive entrepreneurial sub-economies in niche markets, such as Jews and other immigrant groups in North America have done (Light and Bonacich 1988; Portes and Bach 1985). The present immigration wave is too large relative to Israel's population, and is thus likely to be absorbed in the mainstream economy rather than in distinctive niche markets. These immigrants possess substantial human capital combined with great difficulties in advancing as salaried employees, due to present economic realities in Israel. Nevertheless, lack of both tradition of enterprise and extensive ethnic entrepreneurial networks may weaken their prospects of entrepreneurship.

The suitability of the technological incubator project for immigrants from the former USSR has also been debated. These immigrants, while equipped with considerable technological know-how, suffer from a complete lack of the Western marketing approach. By and large, they do not have any experience in forming strategic alliances with business organisations that have the ability to direct their inventions into commercial products and to market them, and they lack any ability to navigate in the complex international markets for technological innovations. It is far from clear whether the assistance offered in most incubators will be able to bridge these gaps. Moreover, the peripheral location of some of the technological incubators may serve to preserve the isolation of new immigrants from major informal business and social networks of technological entrepreneurship. It has also been argued that there is a loss of human capital to the economy when a scientist opens up a grocery store, for example. Often entrepreneurship leads to dead-end, non-challenging jobs with no technological advancement or opportunities. Even the financial rewards are not guaranteed.

A more general controversy is whether entrepreneurship-oriented local development policies, formulated during the early 1990s, do not present another example of a policy adopted too late which better suits the past realities of the 1980s. On the one hand, entrepreneurship is a traditional path for the advancement of immigrants, regardless of economic fluctuations. Moreover, the continuing sluggish performance of the Israeli economy still directs policy-makers into the option of entrepreneurship. On the other hand, it can be argued that economic growth in an era of growing international integration may require far greater emphasis on foreign investment, foreign market penetration, and competition, and on integrating operations within networks of large and small firms. While Israel's political position creates great difficulties in implementing such a development path, promoting small-scale and often unsophisticated entrepreneurship as a major 'remedy' may divert attention from the necessity of overcoming major political-economic obstacles for advancement through the major avenues of growth.

Part IV

RURAL INDUSTRIALISATION IN ISRAEL

16

INDUSTRIALISATION IN RURAL ISRAEL

An introduction

Israeli society is urban in nature. The relatively small rural sector, which accounts for about 8 per cent of the population, is unique and presents us with a variety of interesting experiences of rural industrialisation. The Israeli examples differ greatly from each other and from the experiences of both developing and developed countries. Apart from the traditional Arab sector, which resembles a typical Third World rural periphery in which ownership is private, most Jewish rural settlements exhibit various degrees of collectivisation together with an advanced industrial sector. The distinct duality between the Arab and Jewish rural industrialisation process is part of the overall economic and cultural duality between these two societies within Israel.

This section, therefore, deals with the three major types of rural industrialisation experiences that exist in Israel: rural industrialisation within the kibbutz (plural: kibbutzim) communal settlements, the moshav (plural: moshavim) cooperative settlements, and the traditional Arab non-cooperative settlements. Since kibbutz industries account for the major portion of the industrialisation process in rural areas in Israel, a substantial part of these chapters will be devoted to this sector.

Compared to urban areas, the location of industry in rural areas has only a few absolute advantages, such as an abundance of cheap land or the availability of non-unionised, unskilled labour. Branch plants of multinational companies, that tend to disperse as part of the product cycle process, have accounted for a substantial share of the net growth of manufacturing in rural or non-metropolitan areas of the world. These external initiatives, which encourage labour-intensive, low-paying jobs, in many cases do not contribute to the welfare of rural communities; in fact there is sometimes noticeable exploitation and a lack of long-term commitments to rural areas. In spite of this, many rural regions around the world have attempted to improve their economic situation by encouraging the growth of manufacturing activities in order to establish a sound economy and to discourage the flow of rural–urban migration (Erickson 1976; Lonsdale and Seyler 1979; Whiting 1974).

The uniqueness of Israeli rural industrialisation is that, unlike other countries, it is not geographic and locational considerations that determine the nature and character of the industrialisation processes. Rather, the different socio-cultural, ideological, and organisational factors have fashioned the development of the various distinct societies comprising the rural sector. Any attempt to map rural industries according to industrial branches is bound to fail, therefore, since patterns based on elements of classical industrial location theory do not exist. Two examples of industries that are normally found in or near metropolitan centres should serve to illustrate this point: the sophisticated printing industry, located in remote Kibbutz Be'eri in the Negev Desert; and 'Soft and Easy' – a plant that manufactures a body hair depilatory with world-wide distribution – which is located in the far north of the country in Kibbutz HaGoshrim.

Almost from their inception, Jewish agricultural settlements have been supported by the various Jewish public institutions. Immense prestige was attached to rural settlements. The kibbutz and the moshav combine Zionist pioneering values with socialist values (Etzioni 1957). In the pre-state period, collective settlements constituted the strongest element of the ruling socialist movement. In Zionist ideology, the city has traditionally been looked upon as a necessary evil – a belief that stems mainly from the idea that the country will be built through the conquest of the soil and society rejuvenated by the creation of a healthy peasantry (Cohen 1970). Industries were perceived as part of rural areas, to be combined with agriculture. Some early ideologies used the term 'Agrindus' to describe this integration (Halperin 1963).

In addition to ideology, there were practical reasons to support agricultural settlements: they served as the principal means of increasing Jewish landholdings in Palestine, and thus of broadening the territorial basis of the future state. After the State was founded, these agricultural settlements were perceived as frontier strongholds – a first line of defence – along the hostile borders between Israel and her Arab neighbours (Kimmerling 1983). Furthermore, rural settlements and rural industrialisation policies were part of the continuing overall national goal of population dispersion, which aims at strengthening the geographical periphery and preventing rural–urban migration to the highly populated metropolitan areas of the centre.

The existence of relatively advanced industries in the Jewish rural areas (mainly in the kibbutzim) cannot be explained in terms of branch plants dispersing as part of the well-known product cycle concept, nor can it be attributed only to external government support. The collective organisational structure of these settlements does not leave much room for external initiatives. It is mainly the availability of local entrepreneurs from within, committed ideologically to a collective way of life, which accounts for the success of the industrialisation process. This unique way of life, where collective principles are of equal importance to economic efficiency, has many advantages which, in many cases, contribute to the enhancement of their industries. Location of plants on collective settlements is therefore not determined by objective location factors and profit maximisation considerations. Rather, entrepreneurs in collective settlements are more likely to be born into a location. They are unlikely to question the economic advantage of their place of birth. This strong commitment to the community is a major asset in developing rural industry, and it is sometimes more important than any locational advantage (Bar-El and Felsenstein 1990).

Another major advantage of the collective system for individual industrial entrepreneurs is the fact that they are not risk-takers, as is the case in a free market economy. They are always able to receive the support and backing of their communal settlement and also, in many cases, the backing of the kibbutz movement to which their settlement belongs. This national movement has ensured strong economic and political linkages to the power centres of the country. There have also been less financial constraints on developing industry due to the ample credit supply and internal funding system within the kibbutz movement.

17

THE UNIQUE CASE OF THE KIBBUTZ

THE KIBBUTZ AND ITS BASIC IDEOLOGY

Israel has approximately 290 kibbutzim, varying in population from less than 100 members in the very young kibbutzim, to close to 2,000 in some of the oldest. The average kibbutz, however, has about 440 permanent residents, with the total population of the kibbutz movement in 1990 numbering around 130,000. While this figure represents about 2.8 per cent of the country's population, the kibbutz movement's share in Israel's total civilian labour force constitutes about 5 per cent and accounts for approximately 7 to 9 per cent of Israel's industrial output and about 8 to 9 per cent of industrial exports. This imbalance exists due to the exceedingly high rate of the movement's participation in the labour force – up to 60 per cent of the total kibbutz population (twice as high as the national average) – and the efficiency of its industrial production units.

Kibbutz settlements in Israel evolved as egalitarian communities, with their economies based on diversified agricultural and industrial production. Their physical, social, and economic structures were designed to avoid social differentiation. The guiding principle of the kibbutz movement is best represented by the Marxist definition of ultimate equality amongst members: '[t]o each according to need, from each according to his [or her] ability'. In principle, the kibbutz insisted on complete equality and direct democracy. The ideal of self-sufficiency is reflected in the way most goods and services used by members of the kibbutz are supplied. This applies to education (local schools), food consumption (a common dining hall), clothing, culture and entertainment, housing, medical service, and health care. There is a high degree of homogeneity amongst kibbutz members with respect to values and acceptance of common goals. The average level of formal education of members is two years beyond high school – a level much higher than the national average. Kibbutzim have grown in size, number, and complexity, and today constitute a modern, changing society. The average contemporary kibbutz possesses technically advanced agricultural and industrial branches operated by a skilled labour force. Indeed, the physical, social, and economic structures of the kibbutz are in constant transition, which means that the modern kibbutz shares many characteristics of both urban and rural societies, and may therefore be classified as a 'rurban' community (Kipnis and Salmon 1972; Newman and Applebaum 1989).

Most of Israel's kibbutzim are located in frontier rural areas and possess fairly large farms and agricultural activities as part of their economy, but they are unique and should not be compared to any other form of stereotypical rural community. To appreciate the various aspects of industry in kibbutzim, it is necessary to gain some insight into kibbutz life – into its history and development, and, most of all, into its social values and ideology.

Deganya, established in 1909 near the Sea of Galilee, was the first kibbutz in Israel. The founders of Israel's early kibbutzim played a major role in the country's political and ideological life. Although the kibbutz movement never comprised more than 7 per cent of Israel's total population, there have been times when more than 10 per cent of the Israeli Knesset (parliament) and up to 25 per cent of the cabinet ministers have been kibbutz members (Etzioni 1957). Even today, with Israel's right-wing government in power for more than a decade, the kibbutz movement is overrepresented in the Knesset, and in other major political institutions as well.

In their research on the industrialisation of the kibbutz, Don and Leviatan (1987) list the principles that guide the management of the kibbutz economy – some of which differ significantly from management principles found to exist in other rural societies:

1. Communal ownership of all means of production and all kibbutz property by its members, not as shareholders but as (and only while) members of the kibbutz.
2. Goals of communal as opposed to individual production, and decision-making on a communal rather than on a hierarchical, centralised basis.

3 Communal consumption, which means that the community, through its decision-making mechanisms, determines the distribution of accumulated resources between consumption and investment, and also determines the consumption patterns themselves.
4 Equality in need fulfilment and in effort sharing, and dissociation between effort and material rewards.
5 Implementation of direct democracy by the placement of the supreme authority for all kibbutz matters in the hands of the general assembly, in which every member has one vote.
6 Periodic rotation of one to five years of office holders.
7 A communal system of education and child-rearing, with strong emphasis upon the group as a vehicle for learning.
8 Safeguarding total life security in the livelihood, health, and care of dependants for each member.
9 Maintenance of a social structure that depends only on voluntary cooperation, in accordance with agreed norms.

Organisational characteristics of the kibbutz

The kibbutz is a democratic, self-managed community; as such, most of its domains of activity are assigned to elected committees. Committee activities cover subjects such as economics, education, environmental issues, sports and recreation, social affairs, consumption, health, and the absorption of new members. The coordinators of the major committees, together with several additional elected members, form the secretariat of the kibbutz. The secretariat, which is coordinated by the general secretary, is considered the central elected body of the kibbutz – only the general assembly carries more executive power. Given the large number of committees, about 70 per cent of all kibbutz members will serve in at least one office during a five-year period.

Kibbutz egalitarianism is reflected in the rotation of office holders, with their return to rank and file after one or two periods of service in office. The absence of any enforcement tool in the hands of office holders reflects the voluntary cooperation and intrinsic motivation of kibbutz members. Importance is attached to the needs and capabilities of individual members, and these are taken into account even when forming decisions that concern the kibbutz as a whole. The value placed on communal life is manifested in the emphasis placed on informal relationships. Also stressed is the communication that exists between office holders themselves, and between office holders and the general kibbutz population (Don and Leviatan 1987).

In addition to its internal social and political structure, every kibbutz is affiliated with the highly politicised inter-kibbutz federations. The largest of these, TAKAM, incorporates approximately 60 per cent of all kibbutzim. The second largest federation, Artzi (Hashomer Hatzair), which is more left-wing in orientation, serves around one-third of the general kibbutz population. A smaller federation is Hadati, a religious kibbutz movement that combines socialist communal ideology with an orthodox Jewish lifestyle.

Federations make demands upon the individual kibbutzim for mutual help, as well as for assistance in achieving political and national goals. The headquarters of each federation also assists individual kibbutzim in economic matters, gives advice in matters such as marketing, finance, and supplies, and, when necessary, represents kibbutzim at the governmental level. Each kibbutz is obliged to contribute a certain portion of its manpower (around 5 per cent) to its national organisations, and its income is also taxed progressively, in order to aid young or weak kibbutzim (Don and Leviatan 1987). The internal communal organisation and the external federative umbrella are highly important properties for the understanding of the kibbutz industrialisation process.

Geographical considerations

Most kibbutzim are located in the peripheral regions of Israel. There are historical and ideological bases for this distribution. During Israel's pre-state era, availability of land was a major reason for the site selection of a kibbutz. Other major considerations of the Zionist movement were the occupation and control of territory. After Israel's independence, the location and distribution of new kibbutzim was influenced by national defence considerations, as well as by development priorities; consequently, the largest concentration of kibbutzim is to be found in Israel's northern and southern frontier regions. The peripheral location of most of Israel's kibbutzim has serious implications for the conditions under which kibbutz industries function, since it makes for high communication and transportation costs to Israel's core region, and for the absence of external economies and the agglomeration effects that exist in central areas of the country. Development towns established after 1950 on the periphery often lack the infrastructure at a level necessary to be useful to kibbutzim; local maintenance services, retail trade, banking, and regional cultural institutions are often inadequate for the more sophisticated kibbutz economy and way of life. As a result, the kibbutz has frequently become an island of high-level economic development surrounded by a region with lower overall development (Gradus 1984). The kibbutz population also has a unique ideological, socio-economic, and ethnic character which makes it distinct from its neighbouring non-collective villages and towns. The level of regional economic and cultural integration with these communities is therefore minimal, and interaction between the kibbutzim and their non-collective neighbours is primarily of a technical nature. The kibbutz thus functions within its region without intensive contact with its non-kibbutz environment. In certain regions and times, tensions have arisen; this has occurred especially during government election periods, when kibbutz members tend to vote to the left, whereas residents of small towns in the hinterlands tend to the right.

To compensate for the low level of regional integration, and to overcome the disadvantages of small size and geographical remoteness, kibbutzim have established their own sophisticated country-wide network of service organisations administered through their national organisations. These organisations provide economic assistance to kibbutzim in the form of marketing facilities, auditing, technical and financial consulting, computer services, legal consulting, research, and planning, and sometimes cultural services and educational consulting, as well.

The functional organisation is also reflected in industry. Manufacturing plants are affiliated with the Kibbutz Industry Association (KIA), whose objective is to assist individual plants with external services such as information, market research, financial planning, and technical support, and which has evolved into an efficient and very powerful instrument. Another inter-kibbutz association is the Regional Manufacturing Centre. A country-wide network of these centres performs functions necessary to those individual kibbutzim whose inability to reach economies of scale prevents them from performing such tasks efficiently. For instance, they have large-scale facilities for the packing, sorting, and processing of agricultural produce. They also maintain mechanical equipment and vehicles for regional kibbutzim in large, well-equipped workshops and garages; handle their heavy equipment and laundries; and operate many cultural and educational services, such as schools and cultural events (Bar-On and Niv 1982; Kellerman 1972). Another advantage offered by regional centres is that, unlike the individual kibbutz, they are permitted to hire non-kibbutz labour and can thus operate labour-intensive industries. In this manner, regional centres are able to provide kibbutzim with essential services, and kibbutzim are not forced to compromise their social texture by mixing hired labour with member labour inside their communities. More than 90 per cent of all workers in kibbutz regional centres are hired, mainly from the neighbouring development towns. In contrast, almost all kibbutz members in the centres are employed in managerial positions or in white collar jobs (Don and Leviatan 1987). As might be expected, this division of labour creates another source of regional tension between the residents of the urban development towns and kibbutz members.

INDUSTRIALISATION IN THE KIBBUTZ

Objectives for industrialisation

Four major motives led the kibbutz rural sector to industrialise. First, industrialisation offset the decline in the profitability of agriculture; second, industrialisation allowed the kibbutz to respond to the employment needs of senior members; third, industrialisation created a diversification of employment for female members; and fourth, industrialisation provided professional challenges for the younger generations (Don and Leviatan 1987).

Kibbutzim prospered during the agricultural expansion of the 1950s, when the population of the country tripled due to mass immigration. With the decline in immigration during the 1960s, however, the demand for agricultural produce stabilised, and the only means of maintaining the profitability of agriculture was to mechanise and reduce the labour force. In response to this reduction, an aggressive policy of industrialisation was introduced to the kibbutz sector, with generous government support in the form of subsidised loans and other incentives for industrial projects. An important motive behind the kibbutz sector's programme of industrialisation was the desire to take advantage of the gradual change in relative profitability between the manufacturing industry and agriculture.

Unlike Israel as a whole, there is no official retirement age in kibbutzim. With 90 per cent of all males and 80 per cent of all females above the age of 65 being registered in the active labour force, almost a quarter of the total kibbutz labour force are above 55 years of age (Leviatan 1980b). Consequently, the creation of suitable employment opportunities for senior members of the kibbutz who are no longer able to work in the agricultural sector is a significant challenge to the kibbutz community. Industry has been found to be a reasonably effective solution to the employment needs of the elderly, who are in most cases satisfied with their jobs in industrial plants.

Another challenge posed to the kibbutz community is the creation of fulfilling employment opportunities for female members. During the past decade, over three-quarters of all female members of kibbutzim have been employed either in personal services or in education. This situation has aroused a deep sense of discontent over the failure of kibbutz society to encourage its female members to break away from the traditional housewife–homemaker roles. Industry can offer more diverse opportunities to women who prefer to work in jobs other than personal services and education.

Younger members of kibbutzim also challenge their communities to provide them with fulfilling and progressive work environments. Increasingly, the professional aspirations of the younger generations are associated with high technology. Their desire to join the professional sector makes them distinct from the generation of kibbutz founding fathers, who willingly accepted the necessity of hard and monotonous physical labour as part of their commitment to nation-building. In contrast, many of the younger generation, being born into the ready-made kibbutz, receive professional training in universities, and consequently expect a high level of job satisfaction. High-technology industrialisation has, to an extent, allowed the kibbutz community to respond to the aspirations of this generation, many of whom would otherwise have left the kibbutz to seek jobs in urban areas.

The origin and growth of kibbutz industry

The spatio-temporal evolution

The idea of introducing the manufacturing industry to the kibbutz as a feasible economic venture has always been well rooted in the manifestos of the kibbutz movement. Evidence points to the late 1920s and early 1930s as the embryonic stages of this process. In his research on the origin and diffusion of kibbutz industrialisation,

Meir (1980) divides this more than half-a-century long process into four periods. The 1930s are identified as the first period, when manufacturing shops and small-scale factories were established to supply internal kibbutz needs and limited markets. These factories were engaged in processing agricultural produce and manufacturing basic agricultural equipment. Meir associates the second period of kibbutz industrialisation with the 1940s, when the Second World War was responsible for cutting off the Middle East from traditional supplies of industrial goods from Western Europe. At the same time, the massive presence of the British army and other Allied troops, with their vast financial resources, created great demand for manufactured commodities.

Kibbutzim benefited from the war-induced prosperity; when the war ended, over 100 industrial plants were operating in the kibbutz sector. The war gave rise to the establishment of new food, textile, metal, and packing material plants. Moreover, the post-war period marked the Jewish struggle for independence from the British Mandate and created the need to establish illegal underground munitions workshops. These were transformed into legal factories subsequent to the establishment of the State of Israel in 1948. The third period of kibbutz industrialisation thus began after 1948, and continued throughout the 1950s. This era, considered the 'take-off period', witnessed an enormous increase in Israel's population due to massive immigration. The growing market, along with a shortage of foreign currency, forced the Israeli government to adopt an import substitution policy. Kibbutzim responded intensively to this challenge, establishing food processing plants (though with imported technology).

The fourth and final stage in kibbutz industrialisation, which began in the late 1950s, is associated with a process in which domestic technology and managerial knowledge, as well as information regarding the success of earlier industrial experiences of kibbutzim, were transmitted amongst other kibbutzim. The establishment of the KIA as the consulting agency for kibbutzim in 1963 facilitated the dissemination of such information. From that year on, the process of kibbutz industrialisation has accelerated, both in the number of plants and employees. In addition, the composition of kibbutz industry has shifted from labour-intensive to capital-intensive industries. By the late 1970s, virtually all kibbutzim had already established at least one manufacturing plant.

A spatial analysis of the kibbutz sector's industries reveals a diffusion of economic development from the developed urban and industrial core of the country to its rural hinterland and the frontier periphery. According to Meir's (1980) analysis, industry originated in kibbutzim south of the Tel Aviv metropolitan area – the industrial core of the country – when several kibbutzim developed new plants or converted old workshops into factories in response to the growing demand for manufactured goods in Tel Aviv. Neighbouring kibbutzim, learning about the success of these ventures, adopted this innovation, which was transmitted gradually from the core of the country outward towards the periphery. Meir's analysis reveals a regular pattern of the spatial diffusion of an economic innovation from core to peripheral areas.

The functional evolution

The functional evolution of kibbutz industry is characterised by two major trends which have changed its nature. The first trend is manifested in the movement from labour-intensive to capital-intensive industries. The second is reflected in the transition of market orientation from local use to commercial markets, both national and international, and the upsurge in their participation in competitive markets (Kipnis and Meir 1983). In the early 1930s, during the initial phases of kibbutz development, kibbutz industry was characterised by small workshops engaged in food processing or in the construction of agricultural implements for local markets. The first move towards a commercial industrial system occurred during the 1940s, when food, textile, and metal industries supplied the Allied forces, and later the Israel Defence Forces. During the Second World War, 10 per cent of kibbutzim had only one industrial plant, but by the early 1970s, 78 per cent of kibbutzim had one industrial plant, and over 20 per cent had at least two (Kipnis and Meir 1983) (see Table 17.1).

The early 1970s saw kibbutz industry consolidating previous trends. This period is characterised by Kipnis and Meir (1983) as the 'drive for maturity' stage. It was during this phase that kibbutz industry increased output and productivity, searched for new external markets, and crystallised an industrial branch structure that was compatible to both the inherent communal demands of the kibbutz and to national development policy. During this stage of development, the kibbutz industrial system increased its investments, employment, output, and exports. These developments

Table 17.1 Development of manufacturing plants in kibbutzim 1946-90 (selected years)

Number of	1946	1952	1960	1966	1972	1978	1980	1982	1990
Kibbutzim	101	217	229	230	233	246	255	264	288
Kibbutz plants	126	148	119	148	232	312	315	327	365
Plants per kibbutz	1.25	0.68	0.52	0.65	0.99	1.27	1.24	1.24	1.27
Workers in manufacture	1,153	3,303	5,331	6,980	10,391	13,021	12,455	13,201	17,117
Workers per plant	9.15	22.3	44.8	47.2	44.8	41.7	39.5	40.4	46.9

Sources: Annual Reports, Register of Cooperative Societies; Barkai 1977; *Annual Reports*, Kibbutz Industry Association; Don and Leviatan 1987: 36.

Table 17.2 Distribution of kibbutz industrial activities 1970, 1979, 1988 (in percentages)

Type of Industry	Output			Employees			Establishments		
	1970	1979	1988	1970	1979	1988	1970	1979	1988
Metal products	28.3	24.0	22.9	31.0	28.4	23.8	34.6	26.0	22.0
Plastics	15.4	30.1	30.1	12.4	17.6	25.5	15.6	19.6	20.8
Foodstuffs	16.3	13.2	16.4	17.7	8.8	9.8	10.1	5.7	5.5
Wood and furniture	22.2	10.9	4.7	22.8	16.3	7.4	7.8	5.4	4.5
Electronics	2.4	5.5	4.6	2.9	6.3	6.1	8.4	9.6	7.3

Sources: Kibbutz Industry Association *Annual Reports* 1971, 1980, 1989; Kipnis and Meir 1983: 471.

were achieved by kibbutz industry taking advantage of government incentives and the relatively favourable conditions of European markets. During the late 1970s, kibbutz industry sold 53 per cent of its exports to Europe – mainly to EC countries. The leading exports have been foodstuffs, furniture, metal, plastic, rubber, and electronic products (see Table 17.2).

At the same time, kibbutz industry underwent a structural change. While traditional labour-intensive industries declined, more technologically advanced and capital-intensive industries (such as plastics and electronics) expanded, generating a tendency towards the creation of smaller plants. During the late 1960s, 53 per cent of kibbutz plants employed more than thirty workers; however, by the mid-1970s, such establishments accounted for only 39 per cent of the plants, while those with fewer than ten workers increased from 14 to 24 per cent (Meir 1980). Recently, however, the growth of this group has halted in favour of plants employing ten to thirty workers. This latter group, which accounts for 38 per cent of today's kibbutz plants, has been dominant both relatively and absolutely since the mid-1970s. This implies an emerging pattern in factory size. The reduction in the average workforce per plant has occurred mainly in the traditional labour-intensive industries (see Table 17.3).

These trends reflect the kibbutz community's growing desire to abolish reliance on hired labour on one hand, and the emerging shortage of blue-collar workforce among kibbutz members on the

Table 17.3 Changes in plant size by branch of industry, 1970, 1979, 1988

Type of industry	Average size of plant (no. of workers)		
	1970	*1979*	*1988*
Metal products	43	46	48
Plastics	38	38	55
Foodstuffs	84	65	80
Wood and furniture	138	126	73
Electronics	17	27	37
Other	27	28	31
Total	48	42	43

Source: Kibbutz Industry Association *Annual Reports*, 1971, 1980, 1989.

other. The result has been an increase in productivity, both per employee and per establishment.

The restructuring of kibbutz industry during the 1960s and early 1970s was quite dramatic as compared to changes that occurred in Israeli industry overall. The results of this restructuring were a decrease in industries traditionally operated by kibbutzim and the striking increase in their share of the plastics industry. At present, kibbutzim dominate Israel's plastics industry (see Table 17.4). This dominance is a result of early kibbutz involvement in plastics; it is also a reflection of the way the plastics industry has met the social needs of the kibbutz community. The KIA anticipated the growing world-wide demand for plastic products as substitutes for goods made of metal, wood, and cement. Due to this foresight, they stimulated their affiliated kibbutzim to take advantage of the Capital Investment Encouragement Law and move rapidly into this field. Plastics factories provided diversified indoor job opportunities for the ageing members, as well as for unskilled female members, of the kibbutz. At the same time, this growing industry created new employment opportunities for the younger kibbutz members. The plastics industry did not demand too high a level of scientific expertise, but was relatively sophisticated in its process and met kibbutz professional capabilities and desires.

Types of industry

During the 1960s, the types of kibbutz industry differed significantly from those of Israeli industry as a whole. Industries such as metals, plastics, and food processing – and, to a lesser degree, wood and mining – were over-represented in the kibbutzim. Taken together, they comprised close to 80 per cent of all kibbutz undertakings, compared to just over 60 per cent in the country at large (Don and Leviatan 1987). In contrast, textiles were heavily under-represented, and some major branches, such as diamonds, basic metals, and electronics, were absent altogether. During the early 1960s, the structure of kibbutz industry was similar to that which existed during the Second World War – probably because little had changed in the conditions that guided the industrial policy of the kibbutz. However, composition of kibbutz industry during the early 1970s differs markedly from that existing during the early 1960s.

Characteristic of the 1970s was the heavy commitment of kibbutzim to the plastics industry and the gradual decline in their

Table 17.4 Weight of the kibbutz in the Israeli manufacturing sector, 1960, 1972, 1982 (in per cent)

Type of industry	In employees			In sales (output)		
	1960	1972	1982	1960	1972	1982
Metal and printing	3.6	4.2	3.8	5.2	3.8	4.0
Electronics	–	2.5	2.9	–	2.2	3.2
Wood and wood products	16.8	19.1	10.7	25.3	22.6	9.1
Plastics and rubber	3.9	13.1	24.7	6.4	17.1	37.6
Textiles and leather	1.0	0.7	1.2	1.0	0.6	1.5
Mining and non-metallics	2.7	3.8	3.5	2.4	4.5	1.4
Food	5.2	5.3	3.4	4.7	3.8	2.7
Chemicals and medical	2.8	1.4	2.2	3.3	3.6	0.7
Miscellaneous	6.9	9.4	5.9	13.5	9.3	4.3
Total	3.8	4.3	4.2	4.4	4.2	3.7

Source: Don and Leviatan 1987: 45.

involvement in textiles and food. During that time, kibbutz industry was represented in seventeen of nineteen branches of industry classified by the Israeli Central Bureau of Statistics. Only basic metals and diamonds were excluded (see Figure 17.1).

The 1970s were a more difficult decade than the 1960s for the Israeli economy – and more difficult for kibbutz economy as well. Rates of growth of both employment and production slowed down in the kibbutz manufacturing sector, while investment fluctuated widely from one year to the next. In spite of these setbacks, however, the establishment of new industries continued quite rapidly, emphasising the kibbutz economy's increasing reliance on manu- facturing. During the 1970s, a heavily selective establishment policy on the part of kibbutzim led to substantial changes in the composition of their industries. Probably most noteworthy was the sustained growth of plastics, which became almost entirely associated with kibbutz production during the early 1980s. Other notable developments were the further contraction of the kibbutz wood industry, the continued expansion of industries grouped under the heading 'miscellaneous', and the continued decline in the number of food processing plants in kibbutzim. In many ways, the 1970s and early 1980s were a period of consolidation and readjustment for the kibbutz. Examples of product types according to industrial categories are presented in Table 17.5.

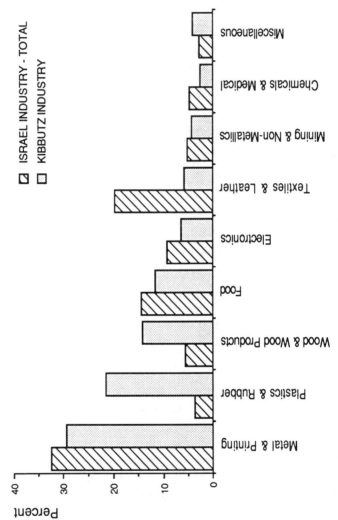

Figure 17.1 Comparison between kibbutz industry and that of the whole of Israel, by major branches

Inter-kibbutz manufacturing partnerships

Economic cooperation between kibbutzim originated in the very early stages of kibbutz development, with inter-kibbutz manufacturing partnerships being only the most recent form of mutual economic ties (Prion 1968; Kellerman 1972). The need for a kibbutz to involve itself in a cooperative economic effort is associated with its unique nature as a collective settlement. Because it is a relatively small and closed socio-economic unit, the kibbutz is unable to

Table 17.5 Examples of product types by branch of industry owned by kibbutzim

Branch of industry	Product type
Metal products	Pressure gauges and regulators, irrigation electronic control systems, solar systems, air conditioning systems, heat exchangers, automatic fire-extinguishing systems, optical microscopes, hydraulic power units, fertiliser spreaders, gas water-heaters
Electrical products	Magnetic cores for transformers, electro-therapy analgesic devices, electric motors, oscilloscopes, computerised monitor systems, television sets, micro-computer systems, communication devices, vacuum equipment, alarm systems, detectors for nuclear radiation
Plastics and rubber	Sprinkler and irrigation accessories, plastic netting, irrigation pipes, drip irrigation products, solar collectors, polyethylene and PVC products, pressure hoses, filters, ftertiliser pumps
Optics	Optical lenses, safety lenses, optical sunglasses, photo-grey lenses, tinted lenses, ophthalmic plastic lenses
Foodstuffs	Spices, fermented products, glucose, starches for food, livestock feed
Building materials	Basalt and granolite aggregates, prepared concrete, bituminous concrete
Wood and furniture	Plywood, chipboard, wooded doors, kitchen furniture, office furniture, modular furniture, children's furniture, wall panels

Sources: Kipnis and Meir 1983: 473; Kibbutz Industry Association *Annual Report* 1982.

mobilise all the human and capital factors of production necessary to fully optimise its operations (Kipnis and Meir 1983). On one hand, the adoption of modern technology complies both with kibbutz aspirations for profitable diversified economies and with membership desires for enriched employment opportunities. On the other hand, modern technology cannot be fully incorporated unless the kibbutz community provides for labour and capital and agrees to take economic risks. Manufacturing partnerships have made it possible for kibbutzim to broaden their range of opportunities for individuals in their communities, as well as to enable them to maximise the potential of their economy. For instance, a kibbutz that owns a manufacturing plant, and whose excess demand is impeded by a shortage of labour, can expand its operations through inter-kibbutz partnerships. Through this means, the kibbutz avoids hiring outside (that is, non-kibbutz) employees – a crucial impediment to the self-employment principles of the kibbutz movement (Kipnis and Meir 1983).

Since the second half of the 1970s, inter-kibbutz partnerships in manufacturing have spread rapidly throughout the kibbutz industrial system. In 1982, forty-one kibbutzim were involved in twenty contracted partnerships; furthermore, some kibbutzim were members of marketing agreements, as well as being involved in various forms of economic partnership arrangements. Usually, inter-kibbutz partnerships include only two or three kibbutzim – oriented as they are to the specific needs of their members—and usually involve one particular economic venture. In many cases, these partnerships are on account of, or in contrast to, the essential interests of the individual kibbutz (Kipnis and Meir 1984). There are a number of prototypes of manufacturing partnerships. We are concerned here with four.

The first prototype is seen in the partnership between two or more kibbutzim, possessing a similar kind of industry, for marketing, purchasing, and/or production inputs. The second is expressed as the partnership between two or more kibbutzim, possessing similar industries, for administration, research and development, and for information and technological exchange. The third is a partnership in a product initially owned by one kibbutz plant and later transferred to another kibbutz as a fully independent line of production; because of a shortage in kibbutz labour, the two independent plants may reach increased total output in a unified production process. A fourth prototype takes the form of a partnership between two kibbutzim or more in a new, jointly initiated manufacturing plant. In most cases, partnerships between kibbutzim are products of a shared ideological system, with the KIA providing the broad framework.

The joint manufacturing ventures between kibbutzim have a variety of location patterns. These patterns reflect the different types of partnerships involved; they also reflect the geographical attributes of distance and space in terms of commuting, marketing, and service considerations. In many cases when commuting is an acute issue, an outlet evolves into an independent plant within the area where the workers reside. Inter-kibbutz manufacturing partnerships have not limited themselves to a given regional context; some partnerships are sub-regional, and nation-wide partnerships exist as well (Kipnis and Meir 1983). However, most are constituted locally and regionally, since their main aim is cost reduction and this can be effected most efficiently on the basis of spatial proximity.

Kibbutz industrial partnerships are a relatively recent phenomenon. They raise some basic issues that might have an impact on kibbutz ideology, as well as on the status of the kibbutz within Israeli society as a whole. Kipnis and Meir (1983) point to many issues concerning the possible and actual consequences of these partnerships. They question whether the kibbutz movement is aware of the socio-economic inequalities that might emerge in the event that some kibbutzim were to enjoy an excess income unshared by other kibbutzim. They also suggest that the existing socio-economic and cultural gaps between the kibbutz and other types of settlements in the same region may deepen. The increase in disparity amongst settlements could be particularly crucial for the moshav and for regional development towns whose workers are deprived of the opportunity to participate in the kibbutz community as hired labour. Another point raised is whether the expanded economies of the kibbutzim, as achieved through their partnerships, will emerge into a new form of regional organisation.

The role of the Kibbutz Industry Association (KIA)

As stated earlier, the disadvantages of the kibbutz in terms of scale economies and geographical remoteness have been partially mediated by the development of national and regional kibbutz associations whose role is to assist the kibbutzim in their planning, purchasing of inputs, marketing, and extension service. The establishment of a national framework – the KIA – in 1963 was a turning point in the planned industrialisation of the kibbutz.

As an interfederative organisation, the KIA enables the small, scattered, independent kibbutz industrial units to operate within a framework of scale economies. It lacks any governing power and its role is restricted to that of coordinator and consultant. In this capacity it functions to promote competitive terms between the kibbutzim and the outside world, while also aiming to reduce internal competition between kibbutzim. This latter objective is extremely important in terms of kibbutz ideology, which regards inter-kibbutz social and economic responsibility as one of its basic principles.

In the economic context of the kibbutz community, the functions of the KIA are multiple. This organisation provides technical services for kibbutzim; finds potential investors and partners in industrial ventures; establishes export outlets and new markets; and checks quality control and standards. In addition to these services, the KIA offers expert tax advice, and organises seminars, courses, and workshops in a variety of fields. The developmental services provided by the KIA consist of furthering the field of research and development, and investigating means to facilitate the proper selection of appropriate industrial enterprises. It is important to note, however, that all services provided by the KIA are offered solely in an advisory capacity; the KIA does not assume any role in the actual production or distribution processes, which remain the responsibility of individual enterprise.

To assist the kibbutz community in maximising the potential of their operations, the KIA has recently established a small research and development institute which studies and conducts research that single, small-sized kibbutz industries lack the resources to undertake (Kipnis and Meir 1983).

Science-based industries and the emergence of robotic technologies

An analysis of the kibbutz industrial structure indicates a relatively low rate of science-based plants. At first glance this seems surprising, since the level of education and technological sophistication among kibbutz members is much higher than the national average. However, when examining the necessary conditions for the establishment of high-technology industries it becomes apparent that the sociological and geographical environment of the kibbutz can act to constrain their development.

Certain conditions are critical for the successful development of high-technology industry: accessibility to scarce information and a skilled labour pool; proximity to research institutes; and a highly advanced scientific infrastructure. In addition, high-amenity areas are needed to make a given location attractive to a highly qualified workforce. There must be a general atmosphere of competition and scientific excitement to stimulate sophisticated production, and external economies of agglomeration, which can be found near large metropolitan areas, are, in most cases, indispensable. Finally, science-based regions are characterised by a process of regular spin-off, whereby new firms are created by entrepreneurs splitting away from parent firms. A complex structure of factors that stimulate and support each other in an upward spiral of industrial growth and product development is therefore necessary to the establishment of most high-technology industries (Christopherson and Gradus 1987).

All or most of these characteristics are absent in the rural kibbutz environment. Indeed, in many cases, these characteristics stand in contradiction to certain kibbutz principles. The rural periphery in Israel, where most of the kibbutzim are located, suffers from limited skilled manpower and a lack of scientific infrastructure. The absence of a tradition of research and development, so necessary to the development of high-technology, is another major barrier to the establishment of these industries. The spin-off effect needed for the expansion of high-technology is not only lacking, but contradicts kibbutz ideology. Kibbutz workers are, in effect, hired for life – they own the plant and cannot be fired. Consequently, they feel strong loyalty to their firms and look forward to a lifetime of job security.

Reluctance to hire outside labour is another major obstacle to the development of science-based industries because specialists in specific, narrow fields do not usually exist among the membership of small kibbutz communities. Kibbutz ideology can cause difficulties even in administration: the principle of rotating management in order to prevent the growth of a management elite prevents specialisation and the accumulation of externalities in terms of information, experience and knowledge so vital to competitive high-technology industries.

Being aware of these disadvantages, the KIA recommends that kibbutzim adopt a policy of developing, strengthening and consolidating those branches of industry where kibbutz society and ideals work to advantage: for example, in the manufacture of food, plastic, metal, agricultural, and irrigation products. Kibbutz industries, because of their specific limitations and constraints, are, therefore, seeking to acquire process technologies suited to their capabilities. These are mainly capital-intensive production systems which use sophisticated machinery, demand relatively few workers, and are characterised by relatively high investment per worker and, therefore, high worker productivity.

Since kibbutzim began switching from agriculture to industry, they have been unable to produce sufficient manpower from their own ranks and have been forced to hire outside labour. Robots are the perfect substitute for hired labour and therefore the rate of the diffusion of robots among kibbutz plants has been outstanding in comparison to the overall rate of robotic development in Israeli industry. While kibbutz industries employ about 5.5 per cent of the total industrial workforce in Israel, they use around 60 per cent of the industrial robots, which are engaged mainly in sorting and packaging agricultural products (Rosner 1988).

The shortage of labour and the limitation in using hired workers are not the only causes of the diffusion of robots; there is also the reluctance to perform routine and alienating tasks and the desire for self-realisation in work – both very important considerations, especially for the more highly educated kibbutz members. Therefore, the introduction of new, advanced technologies has been seen as another way to attract new, educated, young members.

The robotic labour-saving technologies seem custom-made for the unique social structure of the kibbutz. While the introduction of robots in private industry usually provokes strong opposition from unions, in kibbutz industries it has been looked on as a blessing, causing an increase in production and job satisfaction. In many plants, robots have led to an increase in the autonomy of different sub-units and a trend towards decentralisation and increased worker participation (Rosner 1988).

Conflicts between ideology and economic efficiency

The kibbutz industrial system has been a unique and remarkable phenomenon in the Israeli economy. During the last few decades it has spread spatially and functionally throughout Israel's rural sector. This expansion has been responsible for the stimulation of far-reaching modifications to the ideological and socio-economic structure of the kibbutz community.

The industrialisation of the kibbutz has not been achieved without difficulty, however, since this process contains some inherent elements of internal conflict. On the one hand, the kibbutz exists as an economic unit that aims to provide its members with economic security within – and yet separate from – an economic system based on free enterprise. On the other hand, the kibbutz is an ideologically oriented communal settlement that imposes a wide variety of constraints upon its internal economic organisation. For example, industries are often established according to the ideological aspirations of the members of the kibbutz community, which are not always congruent with profit maximisation. Ideology and profitability also conflict in the way kibbutz industry sacrifices a part of its profit for the sake of its commitment to exclude external labour. The size of a plant is therefore limited, since it must be in some feasible proportion to the size of the kibbutz itself.

Kibbutz ideology encounters additional challenges in terms of the management of its industry. Since the general assembly of a kibbutz, which is responsible for kibbutz management at large, functions as, among other things, a board of directors for its industry, it can be viewed as a threat to the egalitarian sensibility of the kibbutz. The same board serves as the ultimate director of other branches of the kibbutz economy as well. Under such circumstances, profit maximisation is not the only factor taken into consideration when formulating decisions, and there is always the danger of conflicts arising amongst the various goals of other economic branches and social needs.

The rotation of managerial posts creates still another conflict between economic efficiency and kibbutz principles. The purpose of this rotation is to preserve equality amongst members by avoiding the consolidation of socio-economic elites within the kibbutz community. The negative effect of this rotation, however, is the way it allows for the erosion of the accumulated managerial experience amongst the executives. The rotation of executives contains certain positive elements as well, since it sustains managerial motivation and encourages new initiatives and ideas.

On a regional level, the kibbutz community faces another challenge to its value system. In order to retain its unique way of life, the kibbutz exists as a relatively closed economic and cultural system – one that is not open to the surrounding non-kibbutz population. But while the kibbutz exists as an isolated entity on the regional scale, it must none the less pursue a balance between its commitments to ideology and its regional integration within a larger environment.

These various conflicts are some of the most crucial ones facing the kibbutz community – conflicts that particularly affect its economic structure. They reflect the unique character of the kibbutz and the discord that exists between its own collectivist commitments and the capitalist structure of Israel's economy. In spite of these constraints, however, kibbutz society has been quite successful in achieving most of the economic objectives it established for itself during its industrialisation process. The relative profitability and efficiency of its industry competes rather successfully with parallel establishments in Israel and abroad. The kibbutzim have learned how to deal with industrial problems within the framework of their socialist ideology, and these objectives are now being achieved to a much higher degree than before. Participation of its workers in the decision-making process has increased; rotation of office holders is an accepted norm; the number of hired labourers has declined; and criteria relating to the quality of working life have become a major concern in kibbutz industry (Don and Leviatan 1987).

Prospects for the future

Past trends and future goals are the basis for speculation on the future of kibbutz industry. The most obvious trend is towards the more technological and capital-intensive industries. Capital to labour ratios, white-collar employees, and the risk factor will increase; and the result of these increases will be the existence of more diversified production lines, both within the kibbutz plant and through inter-kibbutz cooperation. With the globalisation of the world economy, and the need to compete in the world market, KIA externalities may expand to focus on capital mobilisation. They may also concentrate on the provision of research and development and marketing services to the most promising of the growth industries.

Since industry is not restricted to input and output quotas, agriculture will continue to lose its dominant position in the life of the kibbutz (Barkai 1977). This is a change that will result in a significant deviation from the initial philosophy of the kibbutz. Due to the constraints imposed by hired labour restrictions, robotics will become an even more widespread phenomenon. Intensive industrialisation of the kibbutz and increasing cooperation between kibbutzim would widen the gap already prevailing in most regions between kibbutzim and other types of settlements. The latter would experience increasing difficulty in finding jobs when more of their residents who have been hired by kibbutz plants are declared redundant – a process already occurring for ideological and economic reasons. This issue of regional cooperation is so pressing that some kibbutz leaders have suggested that kibbutzim invest in the industrial establishments of the towns within their regions. This is already taking place in the region of Kiryat Shmona in the Galilee. If this process accelerates, existing inequalities in industrialisation between the national core and peripheral regions, as well as within regions, might diminish (Kipnis and Meir 1983).

In view of the globalisation of markets and production, another trend that may be expected is the increase in cooperation between kibbutzim and private entrepreneurs in industrial ventures. This cooperation will be expressed in the establishment of partnerships between kibbutzim and various corporations within Israel and abroad, leading towards the coordination of sales, research and development, global market research, and the exchange of workers, technical personnel, and managerial know-how. This trend may increase the conflict between ideology and profitability within the kibbutz movement, but may narrow the gap between the kibbutz and the outside world, and perhaps lead to a better regional and national integration process.

The arrival of hundreds of thousands of immigrants from the former Soviet Union at the beginning of the 1990s, and the need to absorb them into the society and economy of the country, stimulated the largest kibbutz movement (TAKAM) to adopt a resolution which calls for the absorption of 30,000 new immigrants into their kibbutzim. The idea is to set up government-financed mobile homes in the kibbutzim, where the immigrants would spend up to one year without becoming kibbutz members. They would work for pay in kibbutz enterprises.

Absorption of new immigrants and providing employment for the newcomers has recently become a major goal of Israeli society. Becoming engaged in this process may allow kibbutzim to by-pass ideological constraints on employing externally hired, wage labour. Such arrangements would require departures from kibbutz ideology and may bring a major change in the kibbutz way of life and the nature of its industry. It is too early, however, to speculate on this dramatic event.

18

THE RURAL MOSHAV INDUSTRIALISATION PROCESS

Both the internal and external organisation of moshav rural settlements have been major obstacles to their industrialisation process. Internally, the structure is designed to serve the family farm, the basic agricultural economic unit, which is too small for industrial activities. Externally, the moshav organisations are complex, both functionally and spatially. Ideological and political organisations on national and regional levels distort optimal economic-industrial decision-making since they are not always consistent with the political interests of the moshavim. As a result of the complexity of these organisations, the experience of rural industrialisation in the moshav has been disappointing in comparison with the kibbutz. In order to gain a better understanding of the nature and characteristics of the moshav industry, a short introduction is provided here to the history, ideology and organisation principles of the moshav.

THE ORGANISATION OF THE MOSHAV

The organisational form of the moshav rural settlement was an individualistic reaction within the Zionist socialist movement to the high degree of collectivism and egalitarianism of the kibbutz. In 1921, the first two moshavim – Nahalal and Kfar-Yehezkel – were established in Yizre'el Valley. The principle that the family is the basic unit of production and consumption distinguished these two agricultural settlements from the first kibbutzim founded a decade earlier. Their emphasis was on individual freedom of choice and faith in the effectiveness of economic incentives in guiding individual behaviour (Zussman 1988). In spite of the individualistic nature of the moshav, considerable cooperation has developed in production, marketing and purchasing activities. According to

Etzioni (1957), moshavim, compared with kibbutzim, were considered by the socialist Zionist leaders as second-class members because of the existence of individualistic economic elements in their cooperative structure. Their share in the political elite was, and still is, relatively lower than that of the kibbutz sector. Until the formation of the state, only sixty-four moshavim were established by the Zionist movement in contrast to 137 kibbutzim. In the 1980s, however, the moshav was the predominant form of organisation in Israeli agriculture, with approximately 400 moshavim and about 150,000 people (about 3 per cent of the total population).

The greatest increase in the number of moshavim took place during the mass immigration period of the 1950s. However, these new settlers, most of them new immigrants from Islamic countries, possessed no prior knowledge of the moshav and lacked sympathy for its cooperative principles (Shokied 1971); thus, they became reluctant pioneers (Weingrod 1966). The attempt to overcome their reluctance and socialise them into accepting the moshav principles was usually unsuccessful, and cooperative models of behaviour were frequently enforced from above through economic pressures exerted by the settlement authorities (Schwartz *et al.* 1987).

The founders of the moshav had three main goals: the personal goal of creating secure sources of livelihood; the national goal of participating in the process of returning to the land (mainly in the frontier regions) and thereby changing the traditional Jewish occupational structure; and the social goal of creating a just society based on equality and individual freedom. These goals were guided by several fundamental principles: (a) farming as the main source of income and employment; (b) the family as the basic unit of production; (c) self-employment, which implied no hired labour; (d) mutual aid, which means a high degree of risk-sharing by members of the community; (e) mandatory use of the cooperative communal services such as the purchase of farm inputs, marketing, and provision of health and education services; (f) nationalised land ownership with equitable allocation of land and water; (g) democratic control applied to both the selection of new members and to group choices, with the general assembly as the supreme governing body (Weintraub *et al.* 1969; Zussman 1988).

A superimposed functional and territorial grid of various organisations is one of the major characteristics of the rural settlement system of Israel. A tight network of national and regional entities separates the moshavim from other sectors of the society, creating a major disadvantage for the industrialisation process. While the urban sector is the domain of government agencies, the moshav development process is still largely under the guidance of a quasi-governmental organisation – the Jewish Agency's settlement department. This situation both emphasises and contributes to the maintenance of the cleavage between the rural and urban sectors. In the national sectoral network of organisations, the settlement department of the Jewish Agency was, until recently, the major body responsible for establishing and supporting moshavim. This department has been in charge of locating, planning and establishing new rural settlements, endowing them with appropriate means of production and assisting them until it deems them ready to function independently. The major consequence of this help 'from above' is the formation of long-standing dependency habits among the moshav population which are hardly conducive to industrial entrepreneurship (Bar-El 1984).

Politically, the moshavim, like the kibbutzim, are organised within a framework of national movements, most of which are linked to one of Israel's political parties. The majority of them belong to the 'Tnuat Hamoshavim' (The Moshav Movement), which is associated with Israel's Labour party. The second largest movement is the religious moshav organisation. Tnuat Hamoshavim is deeply split along cleavages of seniority, ethnicity, and regions. Veteran moshavim are located in the better agricultural regions of the country and most of their members are of European and Western origins. The moshavim established after the mass immigration period of the 1950s, however, are often situated in mountainous and/or remote regions on the national periphery of the Galilee and the Negev, and most of their members originate from Middle Eastern and North African countries. The moshav movement has two secretary generals: one representing the veteran European moshavim; and the other the new Oriental moshavim (Gvati 1981).

The main territorial organisation of the moshavim is the regional councils, which are local governments that offer municipal services such as education, supply of water, garbage collection, environment problems, and emergency security services to moshavim within their geographical jurisdiction. In many cases, regional council officials have also lobbied to bring industries into their area so as to provide more employment opportunities to their constituents with jobs. These councils sometimes function jointly with kibbutzim and other types of communities.

A second territorial organisation is the regional supply cooperatives that provide the individual moshav with agricultural supplies on credit at reduced prices. Since the late 1960s, many supply cooperatives have extended their official activities and have become involved in the development of industrial enterprises (Schwartz *et al.* 1987).

The organisational forms of the moshav movement and its ideological principles are the major factors shaping and influencing the industrialisation process in this sector.

INDUSTRIALISATION OF THE MOSHAV

Unlike the kibbutz movement, where information and guidance is available from the KIA, the moshav movement lacks such a central source of business services and basic data to monitor and assist its industrialisation process. This also makes the analysis of the industrial process in the moshav sector more difficult. However, from a limited number of studies, especially that by Schwartz *et al.* (1987), it is possible to obtain a relatively adaquate picture of the major features of moshav industry, its problems, and constraints.

Historically, industrialisation in the moshavim started only in the mid-1960s, more than three decades after the foundation of the first rural kibbutz plants. In the pre-state era, and in the 1950s, the establishment of moshav-based industry was deemed impractical and excited little enthusiasm. It was argued that it is difficult to base an industry upon the nuclear family as a unit of production because industry requires larger production units. Another argument was that the moshav is founded on egalitarian principles, while industry tends to be organised hierarchically, and the organisational structure of the moshav mainly fits the requirements of small units of agricultural production. Furthermore, the economic situation of the veteran moshavim, most of which are in the central portion of the country, was relatively comfortable even without industry. These moshavim managed to reach an acceptable combination of part-time, state-secured and subsidised, self-employment farming supplemented by part-time outside employment in nearby urban centres. Thus, moshav-based industry was only envisaged, and then slowly accepted and adopted – particularly in regions on the remote geographical periphery (mainly in the Galilee) – where no better solution for the employment problem was found (Schwartz *et al.* 1987). The objectives of industrialisation in the moshavim, therefore, is the outcome of two well-known trends in the rural sector: the natural increase of rural population, and the decreasing supply of agricultural jobs due to both severe limitations on agricultural factors of production (mainly water and land) and the continuous improvement of labour-subsituting technologies. According to moshav principles, it is illegal to subdivide the family farm; thus, only one son or daughter may inherit the farm and siblings must look for non-farm employment.

So far, the industrialisation process of the moshav sector has been disappointing. The few plants established in this sector are either agriculture-linked or footloose labour-intensive industries with no need for technical skills. The average wage in these plants is therefore low, and skilled members of the moshavim do not perceive the employment they offer as either attractive or challenging. In a survey conducted in moshavim in the Galilee in the early 1980s by Schwartz *et al.* (1982), of the fourteen plants reported operating in this region, four were owned by individual moshavim, eight by regional organisations, one by the Jewish Agency, and one by the Ministry of Defence. The authors carried out a detailed analysis of eight of these fourteen plants. Some 330 employees worked in the eight plants: less than two-thirds were moshav dwellers; one-fifth were from neighbouring Arab villages; and the remaining 20 per cent included residents of nearby development towns and a few professionals and technicians who commuted from Haifa. The average size for footloose plants was about fifty employees, as compared with around forty workers for agriculture-linked plants (see Table 18.1).

This later survey showed that most employees were male (Schwartz *et al.* 1987). The hiring policy of such plants officially discriminates in favour of moshav members. Manpower stability varies from plant to plant.

Regional ownership seems to be the preferred mode of ownership among the moshav industries, since the assumption is that such a system would facilitate the raising of capital and be able to provide large-scale employment for moshav members, whereas industry as such (that is, plants owned by single moshavim or members of a moshav) might constitute a threat to the internal organisation of the moshav movement. However, a major problem identified with the regional plants is low work discipline, due to the inability to dismiss moshav members. Mixed ownership forms with private ownership from outside the moshav sector seems to be a more promising

Table 18.1 Examples of selected moshav industrial plants in the Galilee by type, product, size, and ownership

Type and product	Number of employees	Ownership
Agriculture-linked plants		
1 Organic mix for growing mushrooms	7	Regional supply cooperatives and the Jewish Agency
2 Chicken and cattle feed	30	Three regional councils
3 Sorting and cold storage of apples and pears	20	The Jewish Agency and several moshavim
4 Sorting and marketing of eggs	70	Several moshavim
5 Sorting, packaging and cold storage of fruit	60	Regional council
Footloose, non-agriculture-linked plants		
1 Production of propylene sacks	55	Single regional council
2 Metal parts for the military industry	60	Single moshav
3 Two production lines: cardboard units for concrete pouring in the construction industry, and plastic trays for meat packaging	40	Koor and several moshavim

Source: Schwartz et al. 1987, pp. 21–2.

model. Such ownership may motivate efficiency and profitability, and it merits further attention. So far, there are few such mixed ownership plants in the Galilee.

The evaluation of the moshav industrial experience does not offer encouraging prospects at this stage. The kibbutz is a much more innovative and highly organised system of rural settlement and can, therefore, overcome the various constraints imposed by ideology. The unstable and political patronage environments which exist in the moshav system, combined with an absence of mutual trust or spirit of cooperation, produce industrial managers who lack professional qualifications or autonomy. They are usually politically constrained in their decisions, particularly with regard to staff, manpower and finance. Kibbutz industry, on the other hand, has been able to reach a significant degree of autonomy in manpower in managerial and even investment decisions. Unlike the moshav, kibbutz industrialisation appears to have achieved its general economic objectives in spite of the various difficulties and constraints.

From this comparison of the kibbutz and moshav industrialisation processes, it is quite clear that the mode of ownership and control has a significant impact on the industrial development process in cooperative rural systems. In the case of moshavim, constitution of the industrial process into a distinct domain separate from the agricultural sector and politics is a prerequisite for successful rural industrialisation (Schwartz *et al.* 1987). The assessment is that until such separation occurs, the interests of industry are likely to be subordinated to agricultural and political considerations, and the results will continue to be disappointing.

19

INDUSTRIALISATION IN THE ARAB SECTOR

RURAL INDUSTRIALISATION IN ARAB VILLAGES

Compared to the Jewish rural sector, the industrialisation process in Israel's Arab rural sector is still in its embryonic stage. This is due mainly to the nature and characteristics of this traditional society, as well as to its complex political situation and the lack of government development policies towards Arab minority regions. Therefore, before dealing with the industrialisation process in the Arab rural sector, a short but essential socio-economic and political introduction is needed.

Arabic-speaking minorities (approximately 880,000 people in 1992), which comprise about 18 per cent of Israel's population, differ in all important cultural parameters from the Jewish population. Between both societies, social separation is considerable. They have separate ethnic origins, speak mutually unintelligible languages, belong to and practise independent religions, and often even consider themselves to be of mutually exclusive nationalities (that is, Palestinian, as opposed to Israeli). Geographically, most of the Arabic-speaking minorities are concentrated in three regions: in the Central Galilee (about 60 per cent); in the 'little triangle' east of Tel Aviv (about 20 per cent); and, in the arid Negev Desert (approximately 8 per cent). These minorities have been marginal in national politics and their representation in national decision-making bodies is minimal. Although formally they are citizens, voters and office holders, and are promised equal rights by law, they have understandable difficulties in sharing the dream of a Jewish national home. Their cultural and religious reference groups and national symbols are located beyond the borders, in countries which are at war with Israel and which often question or deny Israel's legitimacy.

Consequently, Arab-speaking minorities in Israel have been characterised as victims of suspicion by both Israeli Jews and the wider Arab world (Shimshoni 1982; Lustick 1980; Smooha 1980, 1982). The major characteristics of the majority of Arab society in Israel resemble Arab society in the Middle East as a whole: traditional patterns of kinship, extended family structure, and communal subsistence farming. In most cases, Arabs in Israel retain Third World demographic features such as a high birth-rate and a high natural population growth, leading to a low labour participation rate and a consequent high dependency ratio.

Economically, Arab–Jewish dualism was already a feature of the pre-Israel Palestine economy. Dualism is viewed as a non-integrated, parallel development of two interrelated economies within the same territory. The division between these separate economic systems – one, a traditional Arab economy, and the second a modern, Western Jewish economy – has created growing disparities between the two: first under British Palestine, and later in the State of Israel. The Israeli regional development policies towards Arab regions assume that aggregate development projects, no matter where they are placed, will have a 'spread effect' (Myrdal 1957). According to these policies, development projects would generate income and employment throughout the economy, including the Arab regions. The argument has been made that the traditional Arab–Jewish dualism will disappear as the Arab region is integrated with the Jewish economy (Gottheil 1972)

However, since 1948, disparities between Arabs and Jews have tended to increase rather than diminish, which would support Myrdal's hypothesis that unbalanced regional development tends to produce greater imbalances in the economy and among regions, and that the spread effects of development tend to be weak and secondary to the 'backwash effects', which draws resources to the stronger sector in the development process. This argument was supported empirically by Yiftachel (1991) in a recent study dealing with the impact of three new industrial zones (Carmiel, Tefen and Maalot) on Arabs and Jews living in the Galilee. He concludes that economic benefits generated by these industries were disproportionately enjoyed by the Jewish sector, thereby reinforcing patterns of inequality. According to his study, Arabs held only 9 per cent of managerial jobs but 59 per cent of the unskilled jobs. The total spread effects generated by these zones have worked to widen pre-existing economic gaps between Arabs and Jews.

Arab regions were characterised in the 1950s by the predominance of agricultural output and employment and the use of traditional techniques of production. They contrasted sharply with the more advanced and more rapidly expanding Jewish regions. The decline of opportunities in the agricultural sector in the 1960s – due mainly to improvement of labour-saving technologies in the Jewish agricultural sector first and then in the Arab sector – combined with the high natural increase in population in the Arab sector, created a substantial labour surplus in Arab rural villages. The subsequent development of the Jewish sector, especially after the June 1967 War, created opportunities of employment for Arabs outside their villages and within commuting distance.

The emerging picture of the rural Arab situation is, at present, of a mobile and changing society that is increasingly abandoning rural occupations. Only 10 per cent of the Arab rural labour force is employed today in agriculture and two out of three working Arab villagers commute to work in neighbouring cities (Czamanski and Meyer-Brodnitz 1987). This commuting pattern, mainly to low-skilled branches of the Israeli economy such as construction and low-prestige services, may be interpreted as Myrdal's 'backwash effect'. Most of these jobs are found in core regions of the country; that is, the metropolitan areas of Tel Aviv, Jerusalem and Haifa.

The growing surplus labour force, the existing agricultural constraints, and the resulting need to commute to work are ideal conditions to bring about the industrialisation process in Arab villages. However, in spite of the existing potential, there are many obstacles on the road to industrialisation. The local infrastructure within the villages is often unsatisfactory, and remoteness from large urban centres restricts accessibility to important economic services. Other constraints are: the poor quality of the school system, which does not yet adequately provide the skills needed for success in the business world; the lack of access to information; the low level of local institutional development; and the absence of a large class of entrepreneurs, which, together with certain economic institutions, is needed to ensure indigenous development. Economic development is still largely determined by prevailing traditional values such as patterns of kinship and the extended family structure, which are, in most cases, additional obstacles for modernisation and industrial development. The lack of regional economic integration between the Arab rural population and the surrounding Jewish communities, as well as the absence of government policy regarding this issue, leads to deficient participation of rural Arabs in modern economic enterprises (Czamanski and Meyer-Brodnitz 1987).

In contrast to other sectors of the Israeli rural society, mainly kibbutzim and moshavim, there is no evidence that any government effort has been made to introduce industry to the rural Arab sector by providing incentives or by means of other public intervention. In a country whose development is dominated by public policy and strong government intervention, rural Arab villages function in an apparent policy vacuum; therefore, their case represents a unique laboratory in which to observe the workings of the free market forces in determining rates of rural industrialisation in Israel.

In the dual economic system of Israel, the industrialisation process in Arab villages is the product of individual entrepreneurship. These are the risk-takers who combine all the factors of traditional production into new productive organisations. Arab entrepreneurs, few in numbers, are the agents for potential transformation of a traditional rural society into a modern industrial society. The Arab entrepreneurs are in a complex situation, due partly to the newness of this phenomenon and partly to the duality of the society within which this 'development from below' process is taking place.

A survey conducted in the mid-1980s by Czamanski and Meyer-Brodnitz (1987) of the industrialisation of Arab villages found 410 industrial plants employing 8,188 workers. These are mainly non-agricultural, non-service manufacturing activities in very small, workshop enterprises. This activity takes place within more than 100 villages in the Arab–Israeli sector, and the labour force is about 5 per cent of the total Arab–Israeli labour force. It should be stressed that a great proportion of these industrial workers would not qualify for the standard industrial classification of Israel's Bureau of Statistics since they are working in plants that hire less than five employees. However, they perform at least one manufacturing function and are represented as a single separate accounting entity.

According to the survey, 60 per cent of plants employ fewer than 10 workers. There is a very strong sectoral concentration (greater than 80 per cent) in two main labour-intensive industrial branches (sewing and clothing). Eighty-four per cent of the plants are locally owned and 71 per cent of the employees are women. The larger the village, the higher the number of industrial plants and employees in manufacturing. In 40 per cent of Arab villages, mainly the smaller ones, there are no industrial enterprises. The most industrial area in the Arab rural sector is in the upper Galilee due to the fact that there is a high percentage of Druse villages in this region. These villages, because of their more comfortable relations with the Jewish authorities, exhibit a greater propensity towards entrepreneurship than Christian or Muslim Arab villages. Proximity to major urban centres is not a factor determining industrialisation in Arab villages. However, the agglomeration factor is of major importance, since, according to the survey, the chances of establishing new plants are higher in places where other plants already exist.

Geographically, a very noticeable pattern of spatial division of labour is emerging among the rural Arab labourers in Israel. At one end, a female labour force is employed mainly within the traditional villages in labour-intensive branches such as sewing and clothing manufacturing. At the other end, there is the male commuting labour force working in urban metropolitan centres, mainly in construction, services, and labour-intensive manufacturing. It seems that alternative policies of regional development should be pursued by the government in order to change this undesirable spatial division of labour. Development should be directed to Arab villages by providing incentives and opportunities, rather than transforming the Arab population into commuters.

The experience to date shows that industrial entrepreneurship is feasible under free market conditions in the rural Arab sector. This suggests that development efforts should be directed towards the removal of the various obstacles mentioned above.

Part V

POSTSCRIPT

20

ISRAEL'S INDUSTRIAL GEOGRAPHY IN THE 1990s

The previous chapters have outlined major processes shaping Israel's industrial geography during the last two decades. An analysis integrating these processes within alternative scenarios of continued economic stagnation or renewed growth could have served as a solid basis for forecasting future trends and formulating industrialisation plans for the 1990s. However, changing external conditions strongly hint that the 1990s may in some respects present a break from past trends. First, relations between Israel and her neighbouring Arab countries, which have been in a constant turmoil for the last century, are again high on the agenda. Are there any chances for the establishment of political and economic ties between Israel and its neighbours? Will the status of the occupied territories change? Second, a new wave of mass-migration coming from the former USSR (about 350,000 persons in the years 1990 and 1991 alone) has shaken up economic and spatial processes, as well as the perceptions of economic and physical planners.

Finally, the processes that are occurring are largely a consequence of the emergence of a new world order, or 'disorder'. The collapse of the Communist regimes in Eastern Europe, the economic unification of Western Europe, the political and military dominance of the United States versus the increasing economic dominance of East Asia, the shift from the global cold war to regional ethnic conflicts – all these will have a profound impact on Israel's position in the global economy, as well as on its internal space economy. These enormous uncertainties greatly undermine any attempts at outlining future trends. Nevertheless, some prospects are presented below.

POSTSCRIPT

ARABS AND JEWS IN THE OCCUPIED TERRITORIES – ASPECTS OF INDUSTRIALISATION

Industry makes a remarkably small contribution to the GDP of the Arab-Palestinian population in the occupied territories. The proportion of industry even decreased, from 9 per cent of GDP in 1968 to 8 per cent in 1986 (Benvenisti and Khayat 1988). The number of employees in Arab-owned industries in the occupied territories has remained constant since 1970 at about 15,000, a substantial portion of it in olive-oil processing and stone quarries. Arab-Palestinian industry has remained small-scale and traditional, and has hardly restructured since 1967. The industrial production of the occupied territories constituted a mere 1.4 per cent of Israel's industry (Benvenisti and Khayat 1988).

The domestic Arab market has been the major destination for products of Arab-owned industries. Products of these industries sold in Israel have been largely produced by workshops acting as subcontractors to Israeli producers, mainly in textiles and footwear. Economic, administrative, political, and cultural barriers have resulted in the continuation of the backward, underdeveloped nature of industry, despite a relatively rapid rise in the consumption of industrial goods (Bahiri 1987) and the significant number of Arabs from the occupied territories employed in Israeli industry. Industry in the occupied territories has also suffered from Israeli and Jordanian protectionist policies, and from the unequal relationship between the highly developed, state-supported Israeli economy and the backward, traditional Arab-Palestinian economy in the occupied territories (Benvenisti and Khayat 1988). Since 1987, the Intifada (Palestinian uprising) has not produced a favourable atmosphere for industrial development, despite Arab attempts to boycott the purchase of Israeli-made products, the sales of which reached 363 million US dollars in 1984.

Industrial development in Jewish settlements in the occupied territories has accelerated since the early 1980s. Government encouragement consisted of designating all settlements in the occupied territories as 'A' development zones, including those within commuting distance of the Tel Aviv metropolitan area. Moreover, forward provision of industrial buildings was supported in order to attract industries to vacant facilities. These efforts had considerable success, particularly in industrial areas within commuting distance of Tel Aviv. Over 10 per cent of plans for industrial investment, approved between 1984 and 1990 for public support, were in the occupied territories (see Table 20.1). Most of them were rather small and, hence, represented only about 5 to 10 per cent of the total approved investment. Taking into account that some industrial investment was made without public support, particularly in central regions, the relative size of investment in Jewish settlements in the occupied territories is probably smaller still.

Jewish industry in the occupied territories was still significantly

Table 20.1 Israel – approved projects for industrial investment in manufacturing 1984–91 (percentage in development regions)

	Percentage of total number of approved plans	Percentage of total approved investment
Development areas – North [a]		
1984	29.9	28.0
1986	37.1	29.0
1988	37.9	37.0
1990	35.3	42.2
1991	24.5	24.4
Development areas – South [b]		
1984	16.4	14.1
1986	21.8	35.7
1988	18.7	23.7
1990	18.7	25.9
1991	17.5	29.8
Judaea and Samaria [c]		
1984	13.4	7.6
1986	12.7	8.9
1988	15.8	9.3
1990	9.8	4.5
1991	12.0	5.8

Source: Ministry of Industry and Trade, Investment Center Report, various years.

Notes: [a] All localities classified as development zones 'A' or 'B' in northern Israel (Figures 11.1d, 11.1e).
[b] All localities classified as development zones 'A' or 'B' in southern Israel (Figures 11.1d, 11.1e).
[c] Jewish settlements and industrial zones in the occupied territories.

less than 1 per cent of the total Israeli industry during the late 1980s. Nevertheless, in absolute terms, growth was substantial. Barqan industrial zone, near the town of Ariel in western Samaria, has become one of the major magnets for publicly supported industrial investment in Israel. Plants in this industrial zone have full government incentives and enjoy a diversified supply of labour consisting of residents of nearby Jewish settlements, Arabs from nearby localities, as well as commuters from the vast labour market of the Tel Aviv metropolis. The Intifada has not discouraged Jewish industrial investment in the occupied territories. Nevertheless, possible difficulties in enjoying Israel's foreign trade agreements may somewhat discourage exporters from locating there.

Jewish industrial development in the occupied territories has served to strengthen the roots of Jewish settlement in these regions, but is not an essential part of the economy of these settlements. Their economy is based primarily on commuting to the coastal plain and to Jerusalem, since the bulk of the Jewish population lives within commuting distance of the Tel Aviv and Jerusalem metropolitan areas. Hence, economic development in the Jewish settlements in the occupied territories is closely linked with the Israeli system, and practically ignores the completely different Arab economic system within the same region. The future of this dual development, which consists of publicly supported suburbanisation of Jewish industry and Third World small-scale underdeveloped Palestinian-Arab industry, will depend on the political future of the occupied territories and Jewish–Arab relations.

IMMIGRATION AND SPATIAL CHANGE

Mass-migration from the former USSR has generated a new round of comprehensive planning at the national level in an effort to update and revise conceptions of economic-industrial planning as well as of spatial-physical planning. The flow of immigrants has also created spatial shifts originating in market processes; that is, the individual decision-making of immigrants. In terms of spatial policy, the sudden inflow of a large mass of population has led to the re-emergence of calls to utilise the new situation in order to achieve further dispersal of population, a process which practically came to a halt in the 1960s. Old notions emphasising the economic (Pines 1991) and ecological benefits of dispersal, as well as its advantages from the point of view of Israel's security (Gradus 1990a), have been revived.

ISRAEL'S INDUSTRIAL GEOGRAPHY IN THE 1990s

Expanding cities and settlements in the Galilee in order to assure a continued Jewish majority in northern Israel has been strongly advocated (Kipnis 1990; Sofer 1990). Such a strategy is based on the strong preference of the Galilee by the new immigrants themselves and depends on strengthening the economy of metropolitan Haifa. Utilising immigration in order to spawn a renewed growth process in the Negev, which has been stagnant for nearly two decades, has been advocated even more forcefully (Gradus 1990b; Krakover and Stern 1990). The large tracts of land available for immediate development have been a great advantage of the Negev, and have made it a prime area for public housing projects in 1990/1. Development of the Negev has been promoted as a tool to relieve the extreme population crowding of the Tel Aviv metropolis, which presents an extreme security danger in times of war – an old Second World War argument of which Israel received a vivid reminder during the 1991 Gulf War. However, immigrants have shown a low preference for the Negev, largely due to particularly grim employment prospects in the region. Proposed solutions call for the construction of large-scale, publicly supported projects in the south, like the Mediterranean–Dead Sea Canal, a railroad to Eilat, a new international airport near Beer Sheva, and the transfer of military industries to the Negev. None of these projects is regarded by most economists at this stage as cost-effective from the point of view of the Israeli economy as a whole. Nevertheless, a detailed plan, sponsored by the Ministry of Finance, for the settlement of 400,000 immigrants in the Beer Sheva area, has been prepared. This plan calls for massive industrialisation of the area, and its feasibility, with regards to the availability of sufficient financial resources for implementation and the appropriateness of the proposed strategy, is dubious.

Immigrants have tended to prefer locating in the central areas. The existing economic infrastructure and the related employment opportunities have been the major attractions of central areas, particularly the Tel Aviv metropolis. Proximity to relatives and friends have also attracted new immigrants to locations where earlier arrivals have concentrated. It has been argued that calls for the spatial dispersal of immigrants, while in line with practices of the 1950s, do not take into account the vital role supportive ethnic networks and diversified metropolitan opportunities play in the ability of immigrants to search for jobs and create jobs independently of public support (Gonen 1990; Razin 1990c).

In the course of a major effort to revise Israel's national physical statutory outline plan in consonance with the new demographic realities (Lerman and Lerman 1991), the goal of short-term massive population dispersal has been abandoned in favour of short-term concentration and only long-term dispersal. The justification for this decision has been the greater ability to provide employment in the short-term in existing metropolitan areas. Job creation in the periphery has been found to be extremely costly, and the expected multiplier effect of these new costly jobs would have mainly benefited the Tel Aviv metropolitan area. This plan envisages that in the long run, after large-scale public investment in infrastructure in the periphery, economic investment in peripheral regions is expected to be more economical, and even essential, in order to avoid the increasing costs of congestion in the centre.

In practice, the new immigrants have generated a reversal in the process of industrial contraction which characterised the period 1987–9. In addition to expanding local demand, immigrants from the former USSR have particularly penetrated the low-level industrial jobs. Technological incubators are an attempt to exploit the vast human capital embedded in the present wave of immigration. However, technological incubators are marginal in job creation, and a considerable loss of human capital associated with absorbing mass immigration is probably unavoidable. In the short term, large-scale employment can come from full utilisation of hitherto under-utilised capital stock, by encouraging the diffusion of new technologies in existing industry (micro-computer applications, CNC, etc.), and perhaps also by public-sector projects. In the long term, employment growth is likely to be due to new private investment enhanced by the modifications in the Law for Encouraging Capital Investment and the other public measures to promote private investment (see Chapter 11). It is premature to draw firm conclusions about the contribution of this most recent influx of immigrants to structural or spatial change in Israeli industry.

CONCLUSIONS AND FUTURE SCENARIOS

We have come a long way in describing and interpreting the evolution of Israel's industrial geography. Future shifts will depend on the combination of global economic, technological and political trends, as well as on processes observed in Israel in the 1980s and early 1990s, and on government policies. Some remarks concerning the future role of major locational factors are mentioned in this concluding section.

In general, both market mechanisms and shifting government policies suggest that future change in the industrial geography of Israel will consist of gradual dispersal of industry from metropolitan cores to metropolitan fringes or semi-peripheral areas, rather than of leap-frogging the bulk of industrial development into the more remote periphery.

Availability and cost of land cannot be expected to induce rapid dispersal, even under assumptions of most rapid industrial growth. Ample land for industrial development is still available in the fringes of the Tel Aviv metropolis (Lerman and Lerman 1991). The bulk of future land-consuming industrial development is planned to take place in relatively large inter-urban industrial areas (ABC 1991). Nevertheless, plans for such areas are spread throughout the country, including the fringes of the metropolitan areas, and it may take a long time until southern Israel's advantage in the availability of land will become a significant factor in influencing location.

Moreover, real estate considerations can deter the better-established Israeli and foreign corporations from investing in the remote periphery. It has been argued by a large American multinational corporation that a decision to locate a new plant on an expensive tract of land in metropolitan Tel Aviv, rather than to accept the government's offer of free land and a generous incentive package in a remote development town, was well justified in retrospect. In addition to the greater ease of operation in the central location, the value of land in Tel Aviv multiplied several times, whereas the value of land in the peripheral town, subsidised by the government, has remained practically zero. The most profitable location in the near future for stable firms which are not desperate to receive maximal short-term government support, therefore, may well be in industrial areas on the fringes of the Tel Aviv metropolitan area. These areas still offer large tracts of relatively inexpensive land for industrial development which are expected to gain in value as the metropolis expands and non-industrial urban needs begin to compete for land.

The availability and cost of capital is influenced by public policies which usually give preference to remote regions. However, the increasing role of local authorities in economic development efforts may work in favour of semi-peripheral towns which benefit from metropolitan expansion (see Chapter 14). Moreover, even if distortions are eliminated from the government's spatial industrialisation policy (see Chapter 11), it can hardly be expected to succeed in dispersing the more attractive industrial investments beyond semi-peripheral regions within commuting distance of the major metropolitan centres. The immense need for large-scale foreign investment may not allow for the imposition of strict locational criteria upon prospective investors, particularly under the conditions of stiff global competition over investment. The ability of the government to counter market forces in terms of location preferences will be as limited as ever under the expected new global economic order. Encouragement of small businesses (see Chapter 15) may also favour the metropolitan areas, despite public efforts to initiate special schemes for peripheral regions.

Labour costs are not generally a consideration contributing to industrial dispersal in Israel. Labour costs per industrial employee in the Haifa area were 21 per cent higher than the Israeli average in 1988, due to the combination of old large-scale unionised industries and younger high-technology electronics enterprises. In the Tel Aviv subdistrict, labour costs per employee were only 85 per cent of the Israeli average (Central Bureau of Statistics, *Industry and Crafts Survey* 1988). Thus, with the slight exception of Haifa, the Israeli labour market has not produced substantial cost variations due to wage agreements being national in nature and to the availability of cheap Arab labour within commuting distance of nearly any locality in Israel. An exception has been the electronics industry, where a clear core-periphery wage structure, in line with principles of the product cycle model, has emerged. This might have been due to the non-unionised nature of the industry and to its tendency not to employ Arabs. Moreover, it might be based on wage differentials across skill levels, rather than on geographic variations in wage costs of similar skill levels. In general, the semi-peripheral Western Galilee region may become a viable location for industry due to the large concentrations of Russian immigrants in this region, together with the large pool of Israeli Arab labour and relatively easy access to the metropolitan labour market of Haifa.

Agglomeration economies should continue to be a vital factor in the location of specific industrial activities characterised by complex and irregular linkages. If present trends continue, the inner parts of the Tel Aviv metropolis will increasingly specialise in top-level control functions and producer services, as well as in flexible industrial complexes of small-scale enterprises in industries such as fashion, diamonds, etc. Similar complexes in high-technology activities should continue their development in suburban locations. Improvements in *communications* may lead to some dispersal associated with greater spatial specialisation. Utilisation of Israel's advantage in R&D, design centres and advanced production facilities, which operate within global networks of subcontractors and marketing units, will reduce the appeal of Israel's periphery, however. These units depend on extensive international communications, which will tend to override any influence of their modest capital, land, or cheap labour requirements on location decisions. It seems that multinationals may frequently penetrate Israeli high-technology industry through such small-scale advanced operations in the Tel Aviv or Haifa areas. At a later stage, operations may expand into more routine production conducted in semi-peripheral or even peripheral locations.

The more distinct hierarchical spatial division of labour may stress even further the specialisation of the remote periphery in the more routine phases of production. The autonomy of peripheral regions in determining paths of industrial change may further decrease with the increasing global integration of industrial activities. The role of the peripheral regions in networks of multinational corporations and their subcontractors will be specialised, as ever, to routine operations.

A strategy for developing the southern periphery should concentrate on the economic development of Beer Sheva and its surroundings. Decision-makers should consider this as one integrated metropolitan area with shared interests. The concept of Beer Sheva as one regiopolis (a functional city region) was introduced by Gradus and Stern (1980). They argued that this desert region has one unified economic base in which Beer Sheva functions as the service centre of an interconnected urban system which presently contains more than 300,000 people. This structure should dictate future development of the southern periphery as a growth centre. If the goal is to populate Israel's remote desert and provide employment for its inhabitants, emphasis should be put on basic services, infrastructure, and economic benefits capable of competing with those in the centre.

To achieve such a target in the desert periphery, regional development must be planned as one massive unit, rather than as small isolated development towns. Therefore, a compact, functionally interrelated system with a major dominant growth centre capable of providing metropolitan functions and services for the entire regiopolis is desirable. The development of communication and transportation networks to achieve the spread effect from Beer Sheva must be given the highest priority in this commuting region (Gradus and Stern 1980). A good transportation link which will bring metropolitan Beer Sheva closer to Tel Aviv may promote the locational status of the former from peripheral to semi-peripheral and establish Beer Sheva as the fourth metropolitan area of the country.

Most future expansion of the large-scale military industries, as well as the chemical industries, will probably be in southern Israel. However, industrialisation strategies alone do not offer great opportunities for many small and remote towns on the periphery. These will have to pursue alternative strategies, such as the development of tourism and supplying recreational facilities.

Needless to say, any predictions in the Israeli case face great political uncertainties concerning Israel's future boundaries and relations with its neighbours. Any future settlement may influence, for example, the position of Jerusalem, either as a marginal location within Israel or as a hub of Jewish–Arab economic interaction. Future scenarios should take into account alternatives of closed borders versus open borders, and address the issue of the future of the occupied territories, in addition to considering the general assumptions concerning the pace of immigration versus emigration, and of economic growth and structural change.

In July 1992, a new government headed by the Labour Party was formed. Its major theme is a change in national priorities. Indications are that the recommended shift of investment and incentive to the Galilee and the southern Negev regions will now take place with greater intensity, and plans are being presented that would improve the infrastructure of these areas in terms of roads, communications networks, various construction projects, and social services. Further development of Jewish settlement of the occupied territories has already been frozen, which means both a change in emphasis and a diversion of funds. If these policies are successful, this may have an effect on the geographical distribution of populations and industry in Israel.

BIBLIOGRAPHY

ABC (1991) *A General Plan for Industrial Infrastructure in Development Areas*, Submitted to the Ministry of Industry and Trade and to the Ministry of Finance (in Hebrew).
Abramovitch, Z. and Guelfat, I. (1944) *The Arab Economy in Eretz Israel and the Middle East*, Tel Aviv, HaKibbutz HaMeuchad (in Hebrew).
Adizes, I. (1988) *Corporate Life Cycle*, Englewood Cliffs, NJ, Prentice-Hall.
Agassi, J. B. (1980) 'The status of women in kibbutz society', in K. Bartolke, T. Bergman and L. Liegle (eds), *Integrated Cooperatives in the Industrial Society: The Example of the Kibbutz*, pp. 118–30, Assen, Van Gorcum.
Allen, D. N. and Hendrickson-Smith, J. (1986) 'Planning and implementing small business incubators and enterprise support networks', Final report prepared for US Department of Commerce, University Park Pa., Institute of Public Administration, Pennsylvania State University.
Alterman, R. (1988) 'Implementing decentralization for neighborhood regeneration: factors promoting or inhibiting success', *Journal of the American Planning Association*, vol. 54: 454–69.
Amin, A. and Robins, K. (1990) 'The re-emergence of regional economies? The mythical geography of flexible accumulation', *Society and Space*, vol. 8: 7–34.
Amiran, D. H. K. and Shachar, A. (1964) 'The urban geography of Dimona', *Studies in the Geography of Eretz-Israel*, vol. 4: 55–76 (in Hebrew).
Anderson, D. (1982) 'Small industry in developing countries: a discussion of issues', *World Development*, 10 (11): 913–48.
—— and Leiserson, M. W. (1980) 'Rural non-farm employment in developing countries', *Economic Development and Cultural Change*, 28 (2): 227–42.
Applebaum, L. and Newman, D. (1989) *Between Village and Suburb*, Rehovot, Settlement Study Center (in Hebrew).
Atar, D. (1975) 'Gerontological aspects of the kibbutz industry study', *HaKibbutz*, no. 2: 8–63 (in Hebrew).
Avitsur, S. (1976) *Daily Life in Eretz Israel in the XIX Century*, Tel Aviv, Rubinstein and Am Hasefer (in Hebrew).
—— (1985) *Inventors and Adapters*, Tel Aviv, Eretz Israel Museum (in Hebrew).

BIBLIOGRAPHY

Avraham, M. (1985) *With Their Own Resources: A New Look at Investment Required for Economic Development in Israel's Development Towns*, Jerusalem, Jewish Agency, Project Renewal.

Bahiri, S. (1987) *Industrialization in the West Bank and Gaza Strip*, Jerusalem, The West Bank Data Base Project, The Jerusalem Post.

Bamford, J. (1987) 'The development of small firms, the traditional family and agrarian patterns in Italy', in R. Goffee and R. Scase (eds), *Entrepreneurship in Europe*, pp. 12–25, London, Croom Helm.

Bank of Israel (1965) *The Reparations and Their Effect on the Israeli Economy*, Tel Aviv (in Hebrew).

Bar, A. (1990) 'Industry and industrial policy in Israel', in D. Brodet, M. Justman and M. Teubal (eds), *Industrial-Technological Policy for Israel*, pp. 12–25, Jerusalem, The Jerusalem Institute for Israel Studies (in Hebrew).

Bar-El, R. (1984) 'Rural industrialization objectives: the income employment conflict', *World Development*, 12 (2): 129–40.

—— (1985) 'Industrial dispersion as an instrument for the achievement of development goals', *Economic Geography*, 61 (3): 205–22.

—— and Felsenstein, D. (1989) 'Technological profile and industrial structure; implications for the development of industry in peripheral areas', *Regional Studies*, vol. 23: 253–66.

—— —— (1990) 'Entrepreneurship and rural industrialization: comparing urban and rural patterns of locational choice in Israel', *World Development*, vol. 18: 257–67.

——, ——, Bentolila, D., Spiegel, D. and Kedar, F. (1989) *Technological Trends in Rural Industrialization*, Rehovot, Settlement Study Center (in Hebrew).

—— and Nesher, A. (ed.) (1987) *Rural Industrialisation in Israel*, Boulder, Colo. and London, Westview Press.

—— and Schwartz, D. (1985) *Economic Development of Ofakim: A Few Preliminary Considerations*, Rehovot, Settlement Study Center.

——, Schwartz, M., Finkel, R. and Kedar, F. (1982) *Industrialization and Development*, Rehovot, Settlement Study Center (in Hebrew).

Bar-Gal, Y. (1976) 'The industrialisation of the kibbutz as a diffusion process', *HaKibbutz*, no. 3: 221–36 (in Hebrew).

—— and Soffer, A. (1981) *Geographical Change in the Traditional Arab Villages in Northern Israel*, Durham (UK), University of Durham.

Barkai, H. (1977) *Growth Patterns of the Kibbutz Economy*, Amsterdam, North Holland Pub. Co.

—— (1983) 'Theory and praxis of the Histadrut industrial sector', *The Jerusalem Quarterly*, no. 26: 96–108.

Bar-On, D. and Niv, A. (1982) 'The regional development of the kibbutz in the eighties', *Kibbutz Studies*, vol. 7: 2–14.

Beilin, Y. (1987) *Roots of the Israeli Industry*, Jerusalem, Keter (in Hebrew).

Ben Arieh, Y. (1981) 'The development of twelve major settlements in nineteenth century Palestine', *Cathedra*, no. 19: 83–143.

Ben Elia, N. (1975) 'Local entrepreneurship and urban growth under conditions of directed urbanization', MA thesis, Jerusalem, The Hebrew University, Department of Sociology (in Hebrew).

BIBLIOGRAPHY

Ben Moshe, A. (1977) 'Diamond plants in development regions', in *Yearbook of Industry 1977/78*, pp. 12–25, Tel Aviv, Israeli Yearbook Publishers (in Hebrew).
Ben-Porath, Y. (1966) *The Arab Labour Force in Israel*, Jerusalem, Falk Institute (in Hebrew).
—— (1986) *The Israeli Economy, Maturing Through Crises*, Cambridge, Mass., Harvard University Press.
Benvenisti, M. (1976) *Jerusalem, The Torn City*, Minneapolis, Minn., University of Minnesota Press.
—— and Khayat, S. (1988) *The West Bank and Gaza Atlas*, Jerusalem, The West Bank Data Base Project, The Jerusalem Post.
Berman, E. and Halperin, A. (1990) 'Professional labour, defence and growth', in D. Brodet, M. Justman and M. Teubal (eds), *Industrial-Technological Policy for Israel*, pp. 147–82, Jerusalem, The Jerusalem Institute for Israel Studies (in Hebrew).
Berman, M. (1971) 'The location of the diamond-cutting industry', *Annals of the Association of American Geographers*, vol. 61: 316–28.
Biger, G. (1982) 'The spatial distribution of industry in Palestine during the twenties', *Economic Quarterly*, vol. 29: 378–94 (in Hebrew).
—— (1983) 'The industrial structure of towns and counties in Eretz Israel in the 1920s', *Cathedra*, no. 29: 79–112 (in Hebrew).
Biltski, A. (1974) *Solel Boneh 1924–1974*, Tel Aviv, Am-Oved (in Hebrew).
—— (1981) *Creation and Struggle: Haifa Workers' Council 1921–1981*, Tel Aviv, Am-Oved (in Hebrew).
Binah, B. S. (1924) *Industrial Palestine: A Survey of Recent Undertakings and Future Possibilities*, London, W. Speaight & Sons.
Birch, D. (1987) *Job Creation in America: How Our Smallest Companies Put the Most People to Work*, New York, Free Press.
Bluestone, B. and Harrison, B. (1982) *The De-industrialization of America*, New York, Basic Books.
Borukhov, E. (1989) 'Industry in development towns – problems of development', *Economic Quarterly*, 36 (142): 260–71 (in Hebrew).
Boyne, G. A. (1988) 'Politics, unemployment and local economic policies', *Urban Studies*, vol. 25: 474–86.
Bregman, A. (1986) *Industry and Industrial Policy in Israel (1965–1985)*, Jerusalem, Bank of Israel (in Hebrew).
—— (1987) 'Government intervention in industry – the case of Israel', *Journal of Development Economics*, vol. 25: 353–67.
Brock, W. A. and Evans, D. S. (1989) 'Small business economics', *Small Business Economics*, vol. 1: 7–20.
Droido, E. (1946) 'Jewish Palestine. the social fabric', in J. D. Hobman (ed.), *Palestine's Economic Future*, pp. 43–54, London, Percy Lund Humphries.
Bruno, M. (1986) 'External shocks and domestic response: macroeconomic performance, 1965–1982', in Y. Ben-Porath (ed.), *The Israeli Economy, Maturing Through Crises*, pp. 276–301, Cambridge, Mass., Harvard University Press.
Brutzkus, E. (1963) 'Physical planning in Israel', in J. Ronen (ed.), *Israel Economy Theory and Practice*, pp. 203–43, Tel Aviv, Dvir (in Hebrew).

BIBLIOGRAPHY

—— (1964) 'Transformation of the network of urban centres in Israel', *Engineering and Architecture*, 22 (3): 39–47 (in Hebrew).
Campbell, M. (ed.) (1990) *Local Economic Policy*, London, Cassell.
Carmel, A. (1973) *German Settlements in Eretz Israel at the End of the Ottoman Era*, Jerusalem, The Israeli Oriental Association, The Hebrew University (in Hebrew).
Caro, R. A. (1975) *The Power Broker, Robert Moses and the Fall of New York*, New York, Vintage Books.
Castells, M. (1975) 'Immigrant workers and class struggles in advanced capitalism: the Western European experience', *Politics and Society*, vol. 5: 33–66.
Chenery, H., Robinson, S. and Syrquin, M. (1986) *Industrialization and Growth: A Comparative Study*, Oxford, Oxford University Press.
Christopherson, S. and Gradus, Y. (1987) 'High technology in the Holy Land – the origins and consequences of the Israeli development path', in H. Muegge and W.B. Stöhr (eds), *International Economic Restructuring and the Regional Community*, pp. 133–47, Aldershot, Avebury.
Cohen, E. (1970) *The City in the Zionist Ideology*, Jerusalem, The Hebrew University, Institute of Urban and Regional Studies.
—— (1974) 'The power structure of Israeli development towns', in T. N. Clark (ed.), *Comparative Community Politics*, pp. 179–201, New York, Wiley.
Czamanski, D. and Meyer-Brodnitz, M. (1987) 'Industrialization in Arab Villages in Israel', in R. Bar-El and A. Nesher (eds), *Rural Industrialization in Israel*, pp. 143–68, Boulder, Colo. and London, Westview Press.
Dan, H. (1963) *On the Unpaved Road: The Legend of Solel Boneh*, Jerusalem, Schoken (in Hebrew).
Danieli, D. (1958) 'The diamond industry', *Hataassiya*, 22 (5–6): 58 (in Hebrew).
Danielson, M. N. and Doig, J. W. (1982) *New York, the Politics of Urban Regional Development*, Berkeley, Calif., University of California Press.
Deshen, S. (1971) 'Conflict and social change: the case of an Israeli village', *Sociologia Ruralis*, 6 (1): 125–38.
Dicken, P. (1986) *Global Shift*, London, Harper & Row.
Don, Y. (1976) 'Industrializing rural areas: the case of the Israeli moshav', in Y. H. Landau (ed.), *Rural Communities*, New York, Praeger.
—— (1977) 'Industrialisation in advanced rural communities: the Israeli kibbutz', *Sociologia Ruralis*, 17 (1–2): 59–74.
—— and Bar-El, R. (1972) 'Vulnerability of small development towns to depressions', in H. Z. Hirschberg (ed.), *Annual of Bar-Ilan University, Volume in Social Sciences*, pp. 219–46, Ramat-Gan, Bar-Ilan University (in Hebrew).
—— and Leviatan, U. (1987) 'Kibbutz industrialization', in R. Bar-El and A. Nesher (eds), *Rural Industrialization in Israel*, pp. 21–56, Boulder, Colo. and London, Westview Press.
Economic Models (1989) 'Infrastructure 2000 – construction of a freeway network', Ramat-Gan, Economic Models (in Hebrew).

Efrat, E. (1977) 'Industry in Israel's new development towns', *GeoJournal*, 1 (4): 41–6.
—— (1987) *Development Towns in Israel, Past or Future*, Tel Aviv, Achiasaf (in Hebrew).
Eisenstadt, S. N. (1985) *The Transformation of Israeli Society*, London, Weidenfeld & Nicolson.
Elazar, D. J. and Kalchheim, C. (1988) *Local Government in Israel*, Lanham, Md., University Press of America, The Jerusalem Center for Public Affairs.
Eliachar, E. (1979) 'The Palestine government's Census of Industry, 1928', *Economic Quarterly*, vol. 26: 248–56 (in Hebrew).
Erickson, R. S. (1976) 'The "filtering-down" process: industrial location in a non-metropolitan area', *Professional Geographer*, vol. 28: 54–60.
Eshet, G. (1989) 'Rafi Klein, an expensive investor?!', *Mamon*, 13 June: 2–3 (in Hebrew).
Eshkol, L. (1963) 'Development policy in Israel', in J. Ronen, (ed.), *Israel Economy Theory and Practice*, pp. 4–27, Tel Aviv, Dvir (in Hebrew).
Etzioni, A. (1957) 'Agrarianism in Israel's party system', *Canadian Journal of Economic and Political Science*, vol. 23: 363–75.
Evron, I. (1980) *Israel's Defence Industry*, Tel Aviv, Publishing House, Ministry of Defence (in Hebrew).
Fein, A. (1971) 'The kibbutz and problems of agro-industrial integration', in J. Klatzman, B. Ilan and Y. Levi (eds), *The Role of Group Action in the Industrialisation of Rural Areas*, pp. 506–14, New York, Praeger.
Felsenstein, D. (1985) 'An evaluation of the Kiryat Weizmann Science Park', (unpublished report).
—— (1986) *The Spatial Organization of High Technology Industries in Israel*, Jerusalem, Hebrew University, Institute of Urban and Regional Studies.
—— (1988a) *Issues in the Development of High-Technology Industry in Jerusalem*, Jerusalem, The Jerusalem Institute for Israel Studies (in Hebrew).
—— (1988b) 'Spawning the growth process in a non-central location; the case of high technology industry in Jerusalem', *Tijdschrift voor Economische en Sociale Geografie*, vol. 79: 365–75.
—— Janner-Klausner, D. and Wolf, Y. (1991) *Public–Private Partnership in Economic Development*, Rehovot, Development Study Center (in Hebrew).
—— and Schwartz, D. (1991) *The Economic Development of Small Towns Located Within the Orbit of Metropolitan Centres: The Case of Bet-Shemesh and Kiryat-Gat*, Rehovot, Development Study Center (in Hebrew).
—— and Shachar, A. (1988) 'Locational and organizational determinants of R&D employment in high technology firms', *Regional Studies*, vol. 22: 477–86.
Friedland, R. (1984) 'Cashing in on Israel's "chips"; high-technology in the promised land', *Forum*, no. 53: 91–102.
Frohman, D. (1989) 'Organisational culture as a leverage for achieving

BIBLIOGRAPHY

goals of success in science-based industries', *Mashaabei Enosh*, July: 10–15 (in Hebrew).
Gaathon, A. L. (1963) 'Economic planning in Israel', in J. Ronen (ed.), *Israel Economy, Theory and Practice*, pp. 179–202, Tel Aviv, Dvir (in Hebrew).
Giaoutzi, M., Nijkamp, P. and Storey, D. J. (eds) (1988) *Small and Medium Size Enterprises and Regional Development*, London, Routledge.
Gonen, A. (1990) 'Population dispersal and immigrant absorption as conflicting goals', in A. Gonen (ed.), *Geography of Immigrant Absorption, Lessons From the Past and a Look to the Future*, pp. 49–55, Jerusalem, The Israeli Geographical Association (in Hebrew).
Gottheil, F. M. (1972) 'On the economic development of the Arad region in Israel', in M. Curtis and M. Chertoff (eds), *Israel: Social Structure and Change*, pp. 237–48, New Brunswick, N.J., Transaction Press.
Gradus, Y. (1983) 'The role of politics in regional inequality: the Israeli case', *Annals of the Association of American Geographers*, vol. 73: 388–403.
—— (1984) 'The emergence of regionalism in a centralized system: the case of Israel', *Society and Space*, vol. 2: 87–100.
—— (1990a) 'Population dispersal depends on decentralisation of government', in A. Gonen (ed.), *Geography of Immigrant Absorption, Lessons From the Past and a Look to the Future*, pp. 34–5, Jerusalem, The Israeli Geographical Association (in Hebrew).
—— (1990b) 'The shrinking of Israel toward a Jewish Hong Kong in the Middle East', *The Economic Quarterly*, vol. 41: 88–90 (in Hebrew).
—— and Eini, Y. (1981) 'Trends in core-periphery industrialization gaps in Israel', *Geographical Research Forum*, vol. 3: 25–37.
—— and Krakover, S. (1977) 'The effect of government policy on spatial structure of manufacturing in Israel', *Journal of Developing Areas*, vol. 11: 393–409.
—— and Stern, E. (1980) 'Changing strategies of development: toward a regiopolis in the Negev desert', *Journal of the American Planning Association*, vol. 46: 410–23.
Greenwood, N. (1990) 'The nightmares of Israeli small business', Policy Studies, Jerusalem, IASPS, Division for Economic Policy Research.
Greiner L. E. (1972) 'Evolution and revolution as organizations grow', *Harvard Business Review*, 50 (4): 37–46.
Gross, N. (1979) 'Then and now: comments on the Palestine 1928 Census of Industry', *Economic Quarterly*, vol. 26: 257–62 (in Hebrew).
—— (1980) 'Haifa and the early industrialisation of the Yishuv', *Economic Quarterly*, vol. 27: 308–19 (in Hebrew).
Grossman, D., Ben-Basat, R. and Karni, A. (1983) 'Industry at the fringe of the city – the Holon industrial area', in D. Grossman (ed.), *Between Yarkon and Ayalon*, pp. 173–84, Ramat-Gan, Bar-Ilan University Press (in Hebrew).
Gur, B. (1990) 'This sorrow', *Politica*, no. 30: 6–47 (in Hebrew).
Gurevich, D., Gertz, A. and Zankers, A. (1947) *Statistical Handbook of Jewish Palestine, 1947*, Jerusalem, Department of Statistics, The Jewish Agency for Palestine.
Gvati, H. (1981) *Hundred Years of Settlements. The History of Settlements in Israel*, Tel Aviv, HaKibbutz HaMeuchad (in Hebrew).

BIBLIOGRAPHY

Hadar, S. (1984) 'The development of R&D capabilities in Israel: a case study in international development', MA thesis, University Park Pa., University of Pennsylvania, Department of Regional Science.

Haifa and Galilee Research Institute (1989) *The Greater Haifa Bay with a Perspective on the Galilee*, Haifa, University of Haifa.

Halevi, N. and Klinov-Malul, R. (1968) *The Economic Development of Israel*, New York, Praeger.

Hall, P. (1982) *Urban and Regional Planning* (2nd edn), Harmondsworth, Penguin.

—— (1985) 'The geography of the fifth Kondratieff', in P. Hall and A. Markusen (eds), *Silicon Landscapes*, pp. 1–19, Boston, Allen & Unwin.

Halperin, A. (1987) 'Military buildup and economic growth in Israel', *Economic Quarterly*, 33 (131): 990–1010 (in Hebrew).

Halperin, H. (1963) *AGRINDUS: Integration of Agriculture and Industries*, London, Routledge & Kegan Paul.

Hamilton, F. E. I. (1974) 'A view of spatial behaviour, industrial organisations and decision-making', in F. E. I. Hamilton (ed.), *Spatial Perspectives on Industrial Organisation and Decision Making*, pp. 3–43, London, Wiley.

Harris, W. W. (1978) 'War and settlement change: the Golan Heights and the Jordan Rift, 1966–77', *Transactions*, The Institute of British Geographers N.S., vol. 3: 309–30.

Harvey, D. (1989) 'From managerialism to entrepreneurialism: the transformation in urban governance in late capitalism', *Geografiska Annaler*, vol. 71B: 3–17.

Hataassiya (1949) 'Towards the four-year plan, location of industry according to laws of the state or laws of economics', 13 (8-9): 2 (in Hebrew).

—— (1955) 'Is private industry at a standstill?', 19 (8): 1 (in Hebrew).

Horowitz, D. (1946) 'Arab economy in Palestine', in J. B. Hobman (ed.), *Palestine's Economic Future*, pp. 55–65, London, Percy Lund Humphries.

Humphrey, C. R., Erickson, R. A. and Ottensmeyer, E. J. (1988) 'Industrial development groups, organizational resources, and the prospects for effecting growth in local economies', *Growth and Change*, 19 (3): 1–21.

Israeli, O. and Groll S. (1981) 'Implications of an ideological constraint: the case of hired labour in the kibbutz' *Economic Development and Cultural Change*, 29 (2): 341–51.

Jerusalem Institute of Management (1987) *Export Led Growth Strategy for Israel*, Tel Aviv.

Justman, M. (1985) 'Reindustrialisation of the Negev development towns in the 1970s', *Economic Quarterly*, vol. 35: 385–97 (in Hebrew).

Kale, S. R. and Lonsdale, R. E. (1979) 'Factors encouraging and discouraging plant location in nonmetropolitan areas', in R. E. Lonsdale and H. L. Seyler (eds), *Nonmetropolitan Industrialization*, pp. 47–56, Washington, DC, Winston.

Kanovsky, E. (1980) *The Economy of the Israeli Kibbutz*, Cambridge, Mass., Harvard University Press.

Keeble, D., Owens, P. and Thompson, C. (1983) 'The urban–rural

BIBLIOGRAPHY

manufacturing shift in the European community', *Urban Studies*, vol. 20: 405–18.

Kellerman, A. (1972) 'Spatial aspects of the inter-rural centers of Israel', *Journal of Rural Cooperation*, vol. 46: 51–71.

Kibbutz Industry Association. (1971–1989) *Annual Reports*, Tel Aviv (in Hebrew).

Kimmerling, B. (1983) *Zionism and Economy*, Cambridge, Mass., Schenkman.

Kipnis, B. A. (1976) 'Industrial employment in development towns', *Economic Quarterly*, 23 (91): 394–400 (in Hebrew).

—— (1977) 'The impact of factory size on urban growth and development', *Economic Geography*, vol. 53: 295–302.

—— (1990) 'Did we abandon the population dispersal policy?', in A. Gonen (ed.), *Geography of Immigrant Absorption, Lessons From the Past and a Look to the Future*, pp. 17–28, Jerusalem, The Israeli Geographical Association (in Hebrew).

—— and Meir, A. (1982) 'The kibbutz industrial system: a unique industrial community in Israel', in I. F. E. Hamilton and G. Linge (eds), *Spatial Analysis, Industry and the Industrial Environment, Vol. 2*, Chichester, John Wiley & Sons.

—— —— (1983) 'Spatial aspects of inter-kibbutz manufacturing partnerships', *Journal of Rural Cooperation*, vol. 11: 43–64.

—— —— (1984) 'Spatial aspects of inter-kibbutz industrial partnerships', *Economic Quarterly*, 31 (122): 281–89.

—— and Salmon, S. (1972) 'Urban indicators of a kibbutz', *Environmental Planning*, vol. 21–2: 9–21 (in Hebrew).

Kislev, R. (1989) 'The utopia of Hasin-Esh', *Politica*, no. 25: 36–7 (in Hebrew).

Klausner, D. and Shamir-Shinan, L. (1988) *Local Organisation for Economic Development*, Jerusalem, The Jerusalem Institute for Israel Studies (in Hebrew).

Klir, A. (1957) 'Why should we move to Dimona?' *Hataassiya*, 21 (10): 1 (in Hebrew).

Krakover, S. and Stern, E. (1990) 'Application of geographical models for the absorption of the immigration wave from the USSR', in A. Gonen (ed.), *Geography of Immigrant Absorption, Lessons From the Past and a Look to the Future*, pp. 5–16, Jerusalem, The Israeli Geographical Association (in Hebrew).

Lavy, V. (1988) *Unemployment in Israel's Development Towns*, Jerusalem, The Jerusalem Institute for Israel Studies.

Law, C. M. (1985) 'Regional development policies and economic change', in M. Pacione (ed.), *Progress in Industrial Geography*, pp. 219–48, London, Croom Helm.

Lerman, E. and Lerman, R. (1991) *National Outline Plan for Construction, Development and Immigrant Absorption, Interim Report, Stage B*, Tel Aviv, E. and R. Lerman, Architects and Town Planners Ltd (in Hebrew).

Leviatan, U. (1980a) 'Organizational effects of managerial turnover', in U. Leviatan and M. Rosner (eds), *Work and Organization in Kibbutz Industry*, pp. 139–52, Norwood, Pa., Norwood Press.

—— (1980b) 'Work and age: centrality of work in the life of older kibbutz members', in U. Leviatan and M. Rosner (eds), *Work and Organization in Kibbutz Industry*, pp. 43–52, Norwood, Pa., Norwood Press.
—— (1982) 'Higher education in the Israeli kibbutz', *Interchange*, 13 (1): 68–82.
—— (1983) 'Work and aging in the kibbutz: some relevancies for the larger society', *Aging and Work*, vol. 6: 215–26.
—— and Rosner, M. (eds) (1980) *Work and Organization in Kibbutz Industry*, Norwood, Pa., Norwood Press.
Lewin-Epstein, N. and Semyonov, M. (1986) 'Ethnic group mobility in the Israeli labor market', *American Sociological Review*, vol. 51: 342–51.
Light, I. and Bonacich, E. (1988) *Immigrant Entrepreneurs, Koreans in Los Angeles, 1965–1982*, Berkeley, Calif., University of California Press.
Lipshitz, G. (1986) 'Divergence or convergence in regional inequality – consumption variables versus policy variables: the Israeli case', *Geografiska Annaler*, vol. 68B: 13–20.
Liwschitz-Garik, V. (1946) 'The electrification of Palestine', in J. B. Hobman (ed.), *Palestine's Economic Future*, pp. 183–90, London, Percy Lund Humphries.
Lonsdale, R. E. and Seyler, H. L. (eds) (1979) *Nonmetropolitan Industrialization*, Washington, DC, Winston.
Lustick, I. (1980) *Arabs in the Jewish State*, Austin, Tex., University of Texas Press.
Maillat, D. (1988) 'The role of innovative small- and medium-sized enterprises and the revival of traditionally industrial regions', in M. Giaoutzi, P. Nijkamp and D. J. Storey (eds), *Small and Medium Size Enterprises and Regional Development*, pp. 71–84, London, Routledge.
Martinelli, F. (1985) 'Public policy and industrial development in Southern Italy', *International Journal of Urban and Regional Research*, vol. 9: 47–81.
Meir, A. (1979a) 'A dynamic spatial diffusion model: an application to kibbutz industry in Israel', *GeoJournal*, vol. 3: 81–7.
—— (1979b) 'A disparity based diffusion model', *The Professional Geographer*, vol. 31: 382–87.
—— (1980) 'The diffusion of industry adoption by kibbutz rural settlements', *Journal of Developing Areas*, vol. 14: 539–52.
—— (1983) 'Rural industrialization in less developed countries: Review and proposed spatial innovation diffusion approach', *South African Geographer*, 11 (1): 25–39.
Metzer, J. and Kaplan, O. (1985) 'Jointly but separately: Arab–Jewish dualism and economic growth in Mandatory Palestine', *Journal of Economic History*, vol. 45: 327–46.
—— —— (1990) *The Jewish and Arab Economies in Mandatory Palestine: Product, Employment and Growth*, Jerusalem, Falk Institute for Economic Research and Bialik Institute (in Hebrew).
Miller, E. W. (1981) 'Spatial organization of manufacturing in nonmetropolitan Pennsylvania', in J. Rees, G. J. D. Hewings and H. A. Stafford (eds), *Industrial Location and Regional Systems*, Brooklyn, N.Y., J. F. Bergin Publishers.

BIBLIOGRAPHY

Ministry of Commerce and Industry (1960) *Israel's Industrial Future, Outlook 1960–65,* Jerusalem.
—— (1964) *Development Plan of Industry in Israel, Outlook B 1965–1970,* Jerusalem.
Ministry of Industry and Trade (1986) *Objectives for Industrial Development in Israel 1985–1990,* Jerusalem, Center of Industrial Planning (in Hebrew).
—— (1987) *The Israeli Industry 1986,* Jerusalem, Center of Industrial Planning (in Hebrew).
—— (1990) *Towards the 1990s – Structural Issues and Trends of Development in Industry,* Jerusalem, Center for Planning and Economics (in Hebrew).
—— (1991) *Horizons, Business Opportunities, A Share in the Future,* Jerusalem.
Mintz, A. (1983) 'The military-industrial complex: the Israeli case', *Journal of Strategic Studies,* 6 (3): 104–27.
Mokry, B. W. (1988) *Entrepreneurship and Public Policy, Can Government Stimulate Business Startups?,* New York, Quorum Books.
Myrdal, G. (1957) *Economic Theory and Under-Developed Regions,* New York, Harper & Row.
Nathan, R. R., Gass, O. and Creamer, D. (1946) *Palestine: Problem and Promise,* Washington, DC, Public Affairs Press.
Newman, D. (1984) 'Ideological and political influences on Israeli rurban colonization: the West Bank and Galilee mountains', *The Canadian Geographer,* vol. 28: 142–55.
—— and Applebaum, L. (1989) 'Defining the rurban settlement: planning models and functional realities in Israel', *Urban Geography,* 10 (3): 281–95.
Novomeysky, M. (1946) 'The Dead Sea and the world potash industry', in J. B. Hobman (ed.), *Palestine's Economic Future,* pp. 137–45, London, Percy Lund Humphries.
Palestine, Department of Customs, Excise and Trade (1929) *First Census of Industries Taken in 1928 by the Trade Section,* Jerusalem.
Palgi, M. (1980) 'Women members in the kibbutz and participation', in K. Bartolke, T. Bergmann and L. Liegle (eds), *Integrated Cooperatives in the Industrial Society: The Example of the Kibbutz,* pp. 107-17, Assen, Van Gorcum.
Patinkin, D. (1960) *The Israeli Economy in the First Decade,* Jerusalem, Falk Institute for Economic Research (in Hebrew).
Peleg, D. (1980) 'An economic perspective on kibbutz industrialization', in U. Leviatan and M. Rosner (eds), *Work and Organization in Kibbutz Industry,* pp. 7–16, Norwood, Pa., Norwood Press.
Penrose, E. T. (1959) *The Theory of the Growth of the Firm,* New York, John Wiley & Sons.
Pines, D. (1991) 'Population dispersal policy in view of the immigration from the Soviet Union', *Economic Quarterly,* vol. 38: 11–27 (in Hebrew).
Piore, M. J. and Sabel, C. F. (1984) *The Second Industrial Divide,* New York, Basic Books.
Portes, A. and Bach, R. L. (1985) *Latin Journey, Cuban and Mexican*

Immigrants in the United States, Berkeley, Calif., University of California Press.

Portugali, J. (1989) 'Nomad labour: theory and practice in the Israeli–Palestinian case', *Transactions, Institute of British Geographers*, vol. 14: 207–20.

Prion, Y. (1968) *The Planned Development of the Inter-Rural Cooperation in Israel*, Rehovot, The Settlement Study Center, Publication No. 3.

Prister, R. (1987) 'Thorns in the industrial gardens', *Ha'Aretz*, 18 December (in Hebrew).

Rapaport, S. (1984) 'Silicon Valley in Haifa, accelerated growth', *Asakim*, 17 January, pp. 16–17 (in Hebrew).

Razin, E. (1984) *The Location of Industrial Firms in Israel*, Research Paper No. 7, Jerusalem, The Jerusalem Institute for Israel Studies (in Hebrew).

—— (1985) 'The early industrialisation of the Negev, 1950–1960', *Idan*, no. 6: 178–93 (in Hebrew).

—— (1986) 'The effect of organisational structure of industry on the development of peripheral towns in Israel', Ph.D. thesis, Jerusalem, The Hebrew University (in Hebrew).

—— (1988a) 'Ownership structure and linkage patterns of industry in Israel's development towns', *Regional Studies*, vol. 22: 19–31.

—— (1988b) 'The role of ownership characteristics in the industrial development of Israel's peripheral towns', *Environment and Planning A*, vol. 20: 1235–52.

—— (1989) 'Relating theories of entrepreneurship among ethnic groups and entrepreneurship in space – the case of the Jewish population in Israel', *Geografiska Annaler*, vol. 71B: 167–81.

—— (1990a) 'Spatial variations in the Israeli small-business sector: implications for regional development policies', *Regional Studies*, vol. 24: 149–62.

—— (1990b) 'Urban economic development in a period of local initiative: competition among towns in Israel's southern coastal plain', *Urban Studies*, vol. 27: 685–703.

—— (1990c) 'Paths to ownership of small businesses among immigrants in Israeli cities and towns', Paper presented at the 30th Regional Science Association European Congress, Istanbul.

—— (1990d) 'Local opportunities, social networks and entrepreneurship among immigrants: the Israeli experience in an international perspective', Paper presented at the BGU–UCLA Conference on Immigration, Beer-Sheva.

—— (1991a) 'Geographical mobility of selected age cohorts of immigrant groups in Israel: implications for occupational mobility', Presented at the British–Israeli workshop on migration and development, Shefayim (Isr), 3–5 June.

—— (1991b) 'Trends in the industrial geography of Israel during the 1950s and the 1960s', in *Eretz Israel, Vol. 23, David Amiran Volume*, pp. 193–205, Jerusalem, Israel Exploration Society (in Hebrew).

—— and Shachar, A. (1987) 'Ownership of industry and plant stability in Israel's development towns', *Urban Studies*, vol. 24: 296–311.

—— —— (1990) 'The organizational–locational structure of industry in

BIBLIOGRAPHY

Israel and its effects on national spatial policies', *Geography Research Forum*, vol. 10: 1–19.

Reichman, S. (1979) *From Foothold to Settled Territory*, Jerusalem, Yad Ben-Zvi (in Hebrew).

—— and Sonis, M. (1979) 'A master plan for the distribution of Israel's population', *Economic Quarterly*, vol. 26: 55–62 (in Hebrew).

Rivlin, P. (1991) 'Industrial subsidies in Israel', Policy Studies, Jerusalem, IASPS, Division for Economic Policy Research.

Rosner, M. (1988) *High-Tech in Kibbutz Industry: Structural Factors and Social Implications*, Haifa, The University of Haifa, The Kibbutz University Center, Publication No. 73.

Sadan, E. (1976) 'Financial indicators and economic performance in the kibbutz sector', in N. Halevi and Y. Kop (eds), *Israel Economic Papers*, pp. 242–52, Jerusalem, Israel Economic Association and Falk Institute (in Hebrew).

Salomon, I. and Razin, E. (1985) 'Potential impacts of telecommunications on the economic activities in sparsely populated regions', in Y. Gradus (ed.), *Desert Development: Man and Technology in Sparselands*, pp. 218–32, Dordrecht, D. Reidel.

—— —— (1986) *The Geography of the Israeli Telecommunications System: Patterns and Implications*, Jerusalem, The Jerusalem Institute for Israel Studies (in Hebrew).

—— —— (1988) 'The geography of telecommunications systems: the case of Israel's telephone system', *Tijdschrift voor Economische en Sociale Geografie*, vol. 79: 122–34.

Salt, J. (1981) 'International labor migration in Western Europe: a geographical review', in M. M. Kritz, C. B. Keely and S. M. Tomasi (eds), *Global Trends in Migration*, pp. 133–57, New York, Center for Migration Studies.

Schwartz, D. (1985) 'Effective incentive: the Law for Encouraging Capital Investment', *Economic Quarterly*, vol. 32: 12–21 (in Hebrew).

—— (1986) 'The contribution of direct governmental incentives to solving Okakim's employment problems', *Economic Quarterly*, vol. 33: 555–63.

—— (1988) 'The experience of industrialisation in Migal HaEmeq and Maalot', in D. Schwartz and R. Bar-El (eds), *Issues in Regional Development*, pp. 17–46, Rehovot, Settlement Study Center (in Hebrew).

—— (1989) 'The influence of the Law For Encouraging Capital Investment on industrial investment in the development towns', in P. Zussman and Y. Nathan (eds), *Studies in Economics 1988*, pp. 227–50, Jerusalem, The Israeli Association of Economics (in Hebrew).

—— and Felsenstein, D. (1988) *Economic Linkages Between Plants in Development Towns and the Towns: Implications for Local Economic Development*, Rehovot, Settlement Study Center (in Hebrew).

Schwartz, M., Bar-El, R., Finkel, R. and Nesher, A. (1987) *Moshav-Based Industry in Israel*, Working Papers Series 16, Rehovot, Settlement Study Center.

—— Finkel, R. and Bar-El, R. (1982) *Non-Agricultural Employment for Moshav Youth in the Galilee*, Research Report, Rehovot, Settlement Study Center.

BIBLIOGRAPHY

Schweitzer, A. (1984) *Upheavals*, Tel Aviv, Zmora Bitan (in Hebrew).
Scott, A. J. (1988) *Metropolis*, Berkeley, Calif., University of California Press.
—— and Angel, D. (1987) 'The US semiconductor industry: a locational analysis', *Environment and Planning A*, vol. 19: 875–912.
—— and Storper, M. (1987) 'High-technology industry and regional development: a theoretical critique and reconstruction', *International Social Science Journal*, 39 (112): 215–32.
Shachar, A. S. (1971) 'Israel's development towns: evaluation of national urbanization policy', *Journal of the American Institute of Planners*, vol. 37: 362–72.
—— (1974) 'Development processes and spatial structure of the metropolital area of Tel Aviv–Yaffo', *City and Region*, 2 (2): 3–21 (in Hebrew).
—— and Lipshitz, G. (1980) 'Inter-regional migration in Israel', *Studies in the Geography of Israel*, vol. 11: 153–77 (in Hebrew).
Shaliv, A. (1981) *Industry in Israel*, Jerusalem, Ministry of Industry and Trade, Center of Industrial Planning.
Sharon, A. (1951) *Physical Planning in Israel*, Jerusalem, Government Printer (in Hebrew).
Shefer, D. (1972) *The Potential Economic Growth of the Haifa Region, Part A: Survey of Industry in Haifa*, Haifa, Technion Institute for Research and Development (in Hebrew).
—— and Frenkel, A. (1986) *The Effect of Advanced Means of Communication on the Operation and Location of High-Technology Industries in Israel*, Haifa, Technion, Neaman Institute for Advanced Studies in Science and Technology (in Hebrew).
—— —— (1989) *Job Creation in Development Towns in Israel*, Haifa, Technion, Neaman Institute for Advanced Studies in Science and Technology (in Hebrew).
Shimshoni, D. (1982) *Israeli Democracy*, New York, Free Press.
Shinan-Shamir, L. (1984) 'The suitability of industries to non-metropolitan communities: the case of Israeli development towns of Yerucham and Dimona, Ph.D. thesis, Cambridge, Mass., Massachusetts Institute of Technology.
Shokied, M. (1971) *The Dual Heritage: Immigrants from the Atlas Mountains in an Israeli Village*, Manchester, Manchester University Press.
Shtainmets, S. (1989) 'Geographical aspects of R&D-intensive industries in metropolitan Haifa', MA thesis, Haifa, University of Haifa, Department of Geography (in Hebrew).
Smith, N. and Dennis, W. (1987) 'The restructuring of geographical scale: coalescence and fragmentation of the Northern core region', *Economic Geography*, vol. 63: 160–82.
Smooha, S. (1980) 'Existing and alternative policies for Israel's Arabs', *Megamot*, vol. 26: 7–36 (in Hebrew).
—— (1982) 'Existing and alternative policies towards the Arabs in Israel', *Ethnic and Racial Studies*, vol. 5: 71–98.
Sofer, A. (1971) 'The location of industries in the Haifa Bay', Ph.D. thesis, Jerusalem, Hebrew University (in Hebrew).

—— (1976) 'The dispersion of industries in the Haifa Bay', *Studies in the Geography of Israel*, vol. 9: 136–55 (in Hebrew).
—— (1986) 'The territorial conflict in Eretz-Israel', *Horizons, Studies in Geography*, 17–18: 7–23 (in Hebrew).
—— (1989) 'Industry in Haifa during the period of British mandate', *Idan*, no. 12: 147–54 (in Hebrew).
—— (1990) 'Immigrant absorption in northern Israel: possibilities and implications', in A. Gonen (ed.), *Geography of Immigrant Absorption, Lessons From the Past and a Look to the Future*, pp. 36–7, Jerusalem, The Israeli Geographical Association (in Hebrew).
Spiro, M. E. (1968) *Kibbutz Venture in Utopia*, (2nd edn), New York, Schocken.
State Comptroller of Israel (1963) 'Directing the diamond industry to development areas', in *Annual Report No. 13 for 1962*, pp. 181–5, Jerusalem (in Hebrew).
—— (1982) 'Science-based industrial parks', in *Annual Report No. 32 for 1981*, pp. 505–11, Jerusalem (in Hebrew).
—— (1984) 'Aid for establishing electronics plants', in *Annual Report No. 34 for 1983*, pp. 503–8, Jerusalem (in Hebrew).
—— (1988) *Annual Report No. 38 for 1987*, pp. 600–4, Jerusalem (in Hebrew).
—— (1991) *Annual Report No. 41 for 1990*, pp. 517–28, Jerusalem (in Hebrew).
Stöhr, W. B. (1981) 'Development from below: the bottom–up and periphery–inward development paradigm', in W. B. Stöhr and D. R. Taylor (eds), *Development from Above or Below?*, pp. 39–72, Chichester, Wiley.
Storey, D. J. (ed.) (1985) *Small Firms in Regional Economic Development*, Cambridge, Cambridge University Press.
—— (1988) 'The role of small- and medium-sized enterprises in European job creation: key issues for policy and research', in M. Giaoutzi, P. Nijkamp and D. J. Storey (eds), *Small and Medium Size Enterprises and Regional Development*, pp. 140–60, London, Routledge.
Storper, M. and Scott, A. J. (1989) 'The geographical foundations and social regulation of flexible production complexes', in J. Wolch and M. Dear (eds), *The Power of Geography: How Territory Shapes Social Life*, pp. 21–40, Winchester, Mass., Unwin Hyman.
Svirski, S. and Shoshan, M. (1985) *Development Towns: Towards A Different Future*, Haifa, Yated (in Hebrew).
Syrquin, M. (1986) 'Economic growth and structural change: an international perspective', in Y. Ben Porath (ed.), *The Israeli Economy: Maturing Through Crises*, pp. 42–74, Cambridge, Mass., Harvard University Press.
Szenberg, M. (1971) *The Economics of the Israeli Diamond Industry*, New York, Basic Books.
Teubal, M. (1989) 'The future of Israel's sophisticated industry', in P. Zussman and Y. Nathan (eds), *Studies in Economics 1988*, pp. 175–90, Jerusalem, The Israeli Association of Economics (in Hebrew).
Tiger, L. and Shepher, J. (1975) *Women in the Kibbutz*, New York, Harcourt Brace Jovanovich.

BIBLIOGRAPHY

Todtling, F. (1984) 'Organizational characteristics of plants in care and peripheral regions of Austria', *Regional Studies*, 18 (5): 397–412.

Tommel, I. (1987) 'Regional policy in the European Community: its impact on regional policies and public administration in the Mediterranean member states', *Government and Policy*, vol. 5: 369–81.

Toren, B. (1979) 'The incentives for locating textile plants in development towns (1958–1965)', Ph.D. thesis, Jerusalem, The Hebrew University (in Hebrew).

UNIDO (1983) *Industry in a Changing World*, New York, United Nations.

—— (1985) *Industry in the 1980s: Structural Change and Interdevelopment*, New York, United Nations.

Vazquez-Barquero, A. (1987) 'Local development and the regional state in Spain', *Papers of the Regional Science Association*, vol. 61: 65–78.

Watts, H. D. (1981) *The Branch Plant Economy, A Study of External Control*, London, Longman.

Weinblatt, J. and Luski, I. (1986) 'The industrial sector in the Negev during the 1970s: analysis and applications for the 1980s', *Economic Quarterly*, vol. 36: 522–32 (in Hebrew).

Weingrod, A. (1966) *Reluctant Pioneers: Village Development in Israel*, Ithaca, N.Y., Cornell University Press.

Weintraub, D., Lissak, M. and Azmon, Y. (1969) *Moshava, Kibbutz and Moshav: Patterns of Jewish Rural Settlement and Development in Palestine*, Ithaca, N.Y., Cornell University Press.

Weiss, S. (1970) *A Typology of Locally Elected Representatives With Special Reference to the Stability of Local Government in Israel*, Jerusalem, Akademon (in Hebrew).

Weitz, N. and Belkind, S. (1981) 'Leasing industrial land in development areas', *Karka*, no. 20: 25–9 (in Hebrew).

Weitz, R. (ed.) (1989) *The Regional Operation from the Lakhish Region to Region 2000*, Rehovot, Settlement Study Center (in Hebrew).

Whiting, L. R. (1974) *Rural Industrialization: Problems and Potentials*, Ames, Ia., Iowa State University Press.

World Bank (1988) *Social Indicators of Development, 1988*, Baltimore, Md., Johns Hopkins University Press.

Yiftachel, O. (1991) 'Industrial development and Arab–Jewish economic gaps in the Galilee region, Israel', *Professional Geographer*, vol. 43: 163–79.

Zilberberg, R. (1973) *Population Distribution in Israel 1948–1972: Results of Population Dispersal Policy*, Jerusalem, Finance Ministry, Authority for Economic Planning (in Hebrew).

Zussman, P. (1988) *Individual Behavior and Social Choice in a Cooperative Settlement*, Jerusalem, Magnes Press.

STATISTICAL SOURCES

Central Bureau of Statistics, *Census of Population and Housing* 1961, 1972, 1983.

Central Bureau of Statistics, *Industry and Crafts Survey*, various years.

BIBLIOGRAPHY

Central Bureau of Statistics, *Labor Force Survey*, various years.
Central Bureau of Statistics, *Statistical Abstract of Israel*, various years.
Central Bureau of Statistics, unpublished data.
Dun & Bradstreet International, *Dun's 100, Israel's Leading Enterprises*, various years.
Dun & Bradstreet (Israel), *Duns Guide 1991*.
Hatassiya, November 1946, data on the diamond industry.
Ministry of Industry and Trade, yearly accounts of Israeli industry.
Ministry of Industry and Trade, *Investment Center Report*, various years.

INDEX

ABC 231
Abramovitch, Z. 45, 49
Adizes, I. 29
Alterman, R. 176
Amin, A. 9
Amiran, D.H.K. 63
Anderson, D.
Applebaum, L. 174, 175, 187
Arab sector, Arabs and Jews in the occupied territories 226–8: industrialisation in the rural 217–21
Atar, D.
Avitsur, S. 38, 41
Avraham, M. 158, 169, 173, 176
Azmon, Y. 211

Bach, R.L. 75, 179
Bahiri, S. 226
Bamford, J. 177, 179
Bank of Israel 56
Bar, A. 80
Bar-El, R. 22, 24, 68, 75, 95, 149, 161, 174, 185, 211, 212, 213, 214, 215, 216
Bar-On, D. 191
Barkai, H. 52, 117, 195, 208
Barlow Commission report (1940) 51
Beilin, Y. 38, 39, 42, 52, 54, 68
Ben Elia, N. 55, 58
Ben Moshe, A. 64, 65
Ben-Arieh, Y. 36
Ben-Basat, R. 139

Ben-Porath, Y. 9, 12, 28, 39
Bentolila, D. 22, 24
Benvenisti, M. 73, 226
Berman, E. 124, 134, 135, 136
Berman, M. 63, 66
Biger, G. 45, 46
Biltski, A. 43, 52, 56, 57, 69
Binah, B.S. 38
biotechnology industry 136–7, 149
Birch, D. 173
Bluestone, B. 12
Bonacich, E. 179
Borukhov, E. 101
Boyne, G.A. 158
Bregman, A. 17, 96
British Mandate era 38–50; agents of industrialisation during the 41–5; spatial transformation during the 45–50
Brock, W.A. 172
Broido, E. 47
Bruno, M. 12
Brutzkus, E. 51, 52, 56, 67

Campbell, M. 158
capital, access to 14–15
Carmel, A. 38
Caro, R.A. 30
Castells, M. 75
Central Bureau of Statistics 67, 122, 232
Chenery, H. 22
Christopherson, S. 131, 132, 205
climate of Israel 3–4

INDEX

clusters 12, 151; evolution of major high-technology spatial clusters 137–47
Cohen, E. 42, 52, 158, 184
corporate geography 113–23; headquarters locations of large industrial firms 114–16; the role of ownership characteristics 119–21; spatial organisation of the largest multiplant firms 116–19
Creamer, D. 9, 34, 35
cultural differences 131–2
Czamanski, D. 219, 220

Dan, H. 54, 56
Danieli, D. 64
Danielson, M.N. 30
defence industries *see* military industries
Dennis, W. 168
development towns (1956–67) 56–71; the diamond industry 63–6; implications for the large cities 68–71; introduction to the Sapir era 58–63; preconditions for industrial dispersal 56–8; the textiles sector 66–8
development zones 98–112; decentralisation of agencies 109–10; re-evaluation of priorities 110–12; reorientation toward new types of enterprises 109; revision of the map of 103–9
diamond industry 63–6
Dicken, P. 24
dispersal *see* industrial dispersal
diversification 86–97; the case for 91–7; level of 86–90
Doig, J.W. 30
Don, Y. 68, 187, 189, 191, 195, 199, 208
dual economy 38–45

economic change in Israel 8–13
Economic Models 110, 111
Efrat, E. 81
Eini, Y. 74, 84, 86, 90, 95

Elazar, D.J. 158
Eliachar, E. 39
energy sources 14
Erickson, R.S. 184
Eshet, G. 153
Eshkol, L. 52
Etzioni, A. 184, 211
Evans, D.S. 172
Evron, I. 126

Federation of Labour Enterprises (Histadrut) 42, 52; crisis in the 121–3
Felsenstein, D. 74, 110, 125, 126, 127, 129, 130, 131, 132, 139, 142, 146–7, 148, 149, 161, 165, 174, 177, 185
Finkel, R. 95, 211, 213, 214, 215, 216
Frenkel, A. 95, 148
Friedland, R. 129, 130, 133, 135
Frohman, D. 131

Gaathon, A.L. 34
Ganei Taassiya (industrial garden) 149–50
Gass, O. 9, 34, 35
Gertz, A. 41
Giaoutzi, M. 173
Gonen, A. 229
Gottheil, F.M. 218
Gradus, Y. 56, 74, 83, 84, 86, 88, 90, 95, 131, 132, 158, 190, 205, 228, 229, 233, 234
Greenwood, N. 172, 173
Greiner, L.E. 29
Gross, N. 39, 47
Grossman, D. 139
Guelfat, I. 45, 49
Gur, B. 94
Gurevich, D. 41, 44
Gvati, H. 212

Hadar, S. 127
Haifa and Galilee Research Institute 144
Halevi, N. 9, 56
Hall, P. 12, 28, 51, 68
Halperin, A. 124, 127, 134, 135, 136

252

INDEX

Halperin, H. 184
Hamilton, F.E.I. 120
Harris, W.W. 73
Harrison, B. 12
Harvey, D. 158
Hataasiya 54
high-technology industries 12, 124–56, 232; attempts at dispersal 147–52; crisis and restructuring 134–7; evolution of major spatial clusters 137–47; growth of civilian high-technology industry 129–30; growth of military industries 126–9; the origins of 125–6; the role of multinationals 130–4; unselective dispersal 152–6
Horowitz, D. 45

immigration: and ideology 2–3; the local entrepreneurship option 177–80; and spatial change 228–30
incubator parks *see* technology incubators
industrial dispersal: the diamond industry 63–6; implications for the large cities 68–71; incentives for 98–102; introduction to the Sapir era 58–63; level of diversification 86–90; level of industrialisation 83–6; the need for revision 103–12; preconditions for 56–8; stagnation in the periphery 91–7; the textiles sector 66–8 *see also* high-technology industries
industrial geography of Israel: in the 1990s 223–34; background 1–7; during the 1970s–80s 77–180; industry's role in the economy 8–24; and rural industrialisation 181–222; up to 1973 25–76
Industry and Crafts Survey (1988) (CBS) 232
Israeli industry: an international comparative perspective 22–4; changing realities of the 1970s and 1980s 79–82; the development towns (1956–67) 56–71; economic and political change in Israel 8–13; evolution of 27–33; its role in the economy 15–22; major attributes of 15–22; national comparative advantage 13–15; the post-1967 crossroads 72–6; pre-statehood roots 34–5; spatial industrialisation 51–5

Janner-Klausner, D. 176
Jerusalem Institute of Management 132
Jewish Agency's Project Renewal 176, 177
Jewish–Arab economy, dual 38–45; in the occupied territories 226–8
Justman, M. 95

Kalchheim, C. 158
Kaplan, O. 43, 45
Karni, A. 139
Kedar, F. 22, 24, 95
Kellerman, A. 191, 201
Khayat, S. 226
Kibbutz Industry Association (KIA) 190, 193, 195, 196, 198, 201, 203–4
Kibbutzim, industrialisation of the 149, 186–209; basic ideology 186–91; geographical considerations 189–91; ideology and economic efficiency 206–8; inter-kibbutz manufacturing partnerships 201–3; objectives for 191–2; organisational characteristics of the kibbutz 188–9; origin and growth of kibbutz industry 192–8; prospects for the future 208–9; the role of the KIA 203–4; science-based industries and robotic technologies 204–6; types of industry 198–200
Kimmerling, B. 57, 184

253

INDEX

Kipnis, B.A. 91, 95, 96, 187, 194, 196, 201, 202, 203, 204, 209, 229
Kislev, R. 123
Klausner, D. 159, 176
Klinov-Malul, R. 9, 56
Klir, A. 67
Krakover, S. 74, 83, 86, 88, 95, 229

Lavy, V. 162
Law, C.M. 68, 74
Law for Encouraging Capital Investment (1950) 58, 98–112 *passim*, 130, 230
Lerman, E. 230, 231
Lerman, R. 230, 231
Leviatan, U. 187, 189, 191, 195, 198, 199, 208
Lewin-Epstein, N. 75
Light, I. 179
Lipschitz, G. 74
Lissak, M. 211
Liwschitz-Garik, V. 41
local development strategies, re-emergence of 157–68; competing strategies 158–65; factors encouraging the 157–8; the spatial implications 165–8
local entrepreneurship option 169–80; a major route for absorbing immigrants? 177–80; promoting entrepreneurship 174–7; shifting attitudes toward 172–4; trends and spatial variations 170–2
locally owned plants 119–20
Lonsdale, R.E. 184
Luski, I. 95
Lustick, I. 218

Maillat, D. 165
Martinelli, F. 33
Meir, A. 193, 194, 196, 197, 201, 203, 204, 209
Metzer, J. 43, 45
Meyer-Brodnitz, M. 219, 220
military industries 83; the growth of 126–9
Ministry of Commerce and Industry 64, 65, 67, 98

Ministry of Industry and Trade 24, 65, 103, 111, 131, 227
Mintz, A. 127, 128
Mokry, B.W. 179
moshavim 210–16; industrialisation of the 213–16; organisation of the 210–13
multinationals and the role of high-technology 130–4
multiplant firms 116–19, 120–1
Myrdal, G. 218

Nathan, R.R. 9, 34, 35
national comparative advantage 13–15; access to capital 14–15; availability of natural resources 13–14; energy sources 14
natural resources 13–14
Nesher, A. 211, 213, 214, 215, 216
Newman, D. 95, 174, 175, 187
Nijkamp, P. 173
Niv, A. 191
non-local single-plant firms 120
Novomeysky, M. 42

Ottoman era, end of the 36–8
ownership characteristics and industrial development 119–21

Palestine, Department of Customs, Excise and Trade 39
Patinkin, D. 51
Penrose, E.T. 29
physiography of Israel 3–4
Pines, D. 228
Piore, M.J. 8, 173
political change in Israel 8–13
population of Israel 4–6
Portes, A. 75, 177, 179
Portugali, J. 75
pre-statehood roots 34–5; an overview 34–6; the British Mandate era 38–45; end of the Ottoman era 36–8
Prion, Y. 201
Prister, R. 150

Rapaport, S. 144, 147
Razin, E. 52, 64, 65, 67, 68, 70, 73,

75, 76, 92, 93, 100, 102, 108, 110, 114, 116, 118, 119, 148, 155, 156, 158, 160, 161, 165, 171, 172, 173, 177, 229
Regional Manufacturing Centre 190
Reichman, S. 52, 72
Rivlin, P. 99, 100, 110
Robins, K. 9
Robinson, S. 22
Rosner, M. 206
rural industrialisation of Israel 181–221; introduction 183–5; in the Arab sector 217–21; the kibbutzim 186–209; rural moshav industrialisation 210–16
rural-cooperative sector 121

Sabel, C.F. 8, 173
Salmon, S. 187
Salomon, I. 110, 155, 156
Salt, J. 75
Sapir era, introduction to the 58–63, 81
Schwartz, D. 73, 75, 79, 97, 100, 102, 110, 160, 161, 162, 165, 177
Schwartz, M. 95, 211, 212, 214, 215, 216
Schweitzer, A. 57, 65, 81
Scott, A.J. 9, 66, 144, 147
Semyonov, M. 75
Seyler, H.L. 184
Shachar, A. 51, 56, 63, 65, 69, 73, 74, 92, 93, 114, 116, 118, 127, 130, 131
Shaliv, A. 9
Shamir-Shinan, L. 159, 176
Sharon, A. 51
Shefer, D. 69, 95, 148
Shimshoni, D. 218
Shinan-Shamir, L. 66, 75, 76, 91, 93, 94, 96
Shokied, M. 211
Shoshan, M. 80
Shtainmets, S. 144, 149
Smith, N. 168

Smooha, S. 218
Sofer, A. 47, 69, 148, 150, 229
Sonis, M. 72
spatial industrialisation: during the British Mandate era 45–50; early policy (1948–55) 51–5
Spiegel, D. 22, 24
State Comptroller of Israel 64, 102, 137, 152, 153
Stern, E. 56, 229, 233, 234
Stöhr, W.B. 173
Storey, D.J. 172, 173
Storper, M. 9
Svirski, S. 80
Syrquin, M. 8, 22
Szenberg, M. 63, 64, 66

technological incubators 112, 149–50, 179, 230
Teubal, M. 136
textiles sector 66–8
Thompson, C.
Tommel, I. 33
Toren, B. 66
tourism 98, 164
traditional local specialisations 36–8

UNIDO 22, 24

Watts, H.D. 173
Weinblatt, J. 95
Weingrod, A. 211
Weintraub, D. 211
Weiss, S. 158
Weitz, R. 149
Whiting, L.R. 184
Wolf, Y. 176

Yiftachel, O. 76, 218

Zankers, A. 41
Zilberberg, R. 58, 63, 67, 72, 73, 100
Zussman, P. 210, 211